Personal, Social and Emotional Development

with

Understanding the World

and

Mathematics

Brilliant
PUBLICATIONS

Mavis Brown
and Rebecca Taylor

D1341641

Publisher's information

Brilliant Publications
Unit 10
Sparrow Hall Farm
Edlesborough
Dunstable, Bedfordshire
LU6 2ES

Tel: **01525 222292**
Fax: **01525 222720**
e-mail: **info@brilliantpublications.co.uk**
Website: **www.brilliantpublications.co.uk**

Written by Mavis Brown and Rebecca Taylor
Illustrated by Julie Hodgson, Frank Endersby
and Ian Hunt

© Mavis Brown (Personal, Social and Emotional
Development; Knowledge and Understanding of
the World) and Rebecca Taylor (Mathematical
Development).
Revised and updated by Debbie Chalmers 2013

Printed ISBN 978 0 85747 675-3
e-book ISBN 978 0 85747 678-4

First published in the UK in 2013.
10 9 8 7 6 5 4 3 2 1

Printed in the UK.

There are three books in the Foundation
Blocks series, each covering one of the
prime areas and one or two of the specific
areas of the *Early Years Foundation Stage*.
Each book contains a wealth of activities,
ideas and suggestions, clearly described and
illustrated. Further details on how these books
are structured and how they can make it easy
to implement the Department of Education's
new, revised *Statutory Framework for the
EYFS* (September 2012) are given in the
Introduction.

The other books in the Foundation Blocks
series are:
*Communication and Language with Literacy
by Irene Yates*
Encourages children to speak and listen,
express themselves clearly, use language
confidently, use and enjoy books, link sounds
and letters, develop phonic knowledge and
begin to read and write.

*Physical Development with Expressive Arts &
Design
by Mavis Brown and Maureen Warner*
Encourages children to be active, move with
confidence, control and coordination, make
healthy lifestyle and food choices and learn to
care for personal needs. It also encourages
children to explore media and materials, use
imagination in design and technology and
share thoughts, ideas and feelings through
dance, movement, art, music and role-play.

Contents

© Mavis Brown and Rebecca Taylor
and Brilliant Publications

Introduction

This book has over 160 differentiated activities set in real-life contexts relevant to children in the Early Years Foundation Stage (EYFS). The activities aim to develop children's skills and to meet the Department of Education's learning and development requirements in the areas of *Personal, Social and Emotional Development, Understanding the World* and *Mathematics*. They offer opportunities for practitioners to follow the guidance document *Development Matters* and to encourage and support children as they work towards the Early Learning Goals (ELGs) of the new, revised Statutory Framework for the Early Years Foundation Stage (September 2012).

There are three ELGs for the prime area *Personal, Social and Emotional Development*: Self-confidence and Self-awareness; Managing Feelings and Behaviour; and Making Relationships.

A table showing which learning opportunities are addressed by each activity can be found on pages 207–217.

Children who feel positive about themselves and their individuality are more likely to be independent and better emotionally adjusted and to approach life with a positive attitude. Those who work cooperatively and form constructive relationships are more likely to become both academically and personally successful. They will also be able to understand and explore issues of right and wrong and stand against discrimination and social injustice.

Practitioners must provide an environment that encourages children to develop independence and self-confidence. They should be willing to try new activities and able to say why they like some activities more than others. They need to practise speaking confidently within a familiar group of adults and peers, talking about their own ideas and choosing the resources they need for different projects. They must understand how and when to ask for help or accept support, but also enjoy working and playing independently, alone and with other children.

Adults should act as good role models, demonstrating to children how to work as members of a group or team. Children need to learn how to understand and follow rules when appropriate. They should be encouraged to talk about how people show feelings, about the importance of trust, honesty and reliability, and about acceptable and unacceptable behaviours and their consequences. As they mature, they should learn to adapt their behaviour to different situations and to accept changes of routine without distress or loss of confidence.

It is equally important that children can play cooperatively, sharing, taking turns, solving problems and discussing and respecting the ideas and opinions of others. They need to learn to be sensitive to others' feelings and to be able to form positive relationships with adults and with other children.

There are three ELGs for the specific area *Understanding the World*: People and Communities; The World; and Technology.

Adults can help children to develop a positive self-image and a sense of belonging, as well as respect for other people, through exploring similarities, differences and attitudes within cultures and communities and by modelling promoting equal opportunities and high self-esteem for everybody.

Children should be encouraged to talk about past, present and future events in their lives and the lives of their families. With support, they can learn to express their feelings about personal events and their likes, dislikes and opinions, while understanding that people don't always enjoy the same things and that different cultures, beliefs and traditions must be respected. Practitioners should always help children to celebrate diversity.

It is important to provide an environment that allows children to learn through their senses and through first hand experiences. Children must be encouraged to display curiosity and to observe and explore in order to discover the answers to their questions. They need to lean to predict, experiment, think and make decisions for themselves, with support from adults where appropriate.

Practitioners must draw children's attention to the features of their environment and explain how environments can vary. They should provide opportunities to use scientific equipment and to explore similarities and differences between places, objects, materials, plants and animals. Children need to talk about how and why things happen and the changes that they observe, as well as how to care for their environment.

Children should recognize that a range of technology is used in the world today. Practitioners must offer opportunities and support for children so that they may learn to select and use technology for particular purposes, both in their early years setting and at home. A range of simple programs are available to help children to learn to use ICT equipment, such as a computer screen, mouse and keyboard. They should also have

© Mavis Brown and Rebecca Taylor
and Brilliant Publications

access to programmable toys, such as 'Roamers' and 'Bee-bots', and to electronic play equipment, such as hand held games and toys operated by battery or remote control.

In role-play, children can learn about technological equipment by using pretend telephones, walkie-talkies, cash tills and kitchen appliances. They may also be introduced to the uses of technology in everyday life, through participating in adult-led activities involving cookers, microwaves, toasters, blenders, food mixers, printers, photocopiers, laminators, etc. Adults should provide appropriate additional equipment, such as cameras, CD players and musical keyboards, for children to use whenever they wish to, with supervision.

There are two ELGs for the specific area *Mathematics*: Numbers; and Shape, Space and Measures.

Children should develop the ability to count reliably, using numbers 1–20. They need to learn not only to say the number names in order, but also to match the numbers to objects counted, know that you say one number for each object you count and that, when you count, the last number you say gives you the number of objects in the group.

Practitioners should support children in learning to say which number is one more or less, and to add and subtract single digit numbers, by counting on or back to find the answer. They need to provide practical opportunities for calculating and talking about numbers in everyday situations. Children begin to develop an understanding of addition, subtraction, multiplication and division through comparing and combining numbers of objects, as well as through adding groups of the same number of objects and sharing objects equally between some children.

It is important that children also learn to recognize and write numerals correctly and understand that they can be used as labels for house numbers, bus routes and television channels, or combined to create telephone numbers or references for items ordered or bookings made.

A variety of activities can help children to develop an understanding of the properties of both 2D and 3D shapes and the ability to describe the characteristics of objects using mathematical language. They should be encouraged to identify and name shapes in the environment and to develop spatial awareness through handling shapes and fitting them together. Comparing and ordering items by size, length and capacity is also important, using both standard and non-standard units of measurement.

Daily routines and related activities will help children to develop an understanding of time and to learn the days of the week and the months of the year. Adults should use everyday language to talk with children about size, weight, length, capacity, position, distance, time and money, to compare quantities and objects and to solve problems. They must also provide opportunities for children to recognize patterns and explore mathematical concepts.

The EYFS Statutory Framework states that children learn most effectively by investigating, experiencing and 'having a go', concentrating, trying again and enjoying their own achievements, creating, linking ideas, thinking critically and developing their own strategies for doing things.

Most of the activities in this book can be adapted to fit any topic or project that is being explored by a group of children. The tasks are designed to allow practitioners to choose their own directions and to include opportunities for exploration and new challenges as children's skills are being developed.

The activities are divided into eight sections to indicate which prime or specific area and which of the ELGs each one will particularly contribute to. Practitioners may use this information when planning, to ensure that they are not concentrating too much on one area at the expense of another. They should be providing enough encouragement and experiences for all children to speak and play confidently, manage behaviour, form positive relationships, respect other people, observe features in the environment, use some basic technology, solve number problems and explore mathematical concepts.

Some activities contribute to more than one ELG and many are also strongly linked to other learning and development areas and their ELGs. These are listed under the pages headed 'Table of learning opportunities' (pages 207–217).

To avoid the clumsy 'he/she', the child is referred to throughout the book as 'she'.

Personal, Social and Emotional Development with Understanding the World and Mathematics

Planning

Where relevant, the activities have been linked to sixteen popular topics that are frequently used in early years and primary settings. These are:

- Animals
- Colours
- Food and shopping
- Health
- Myself
- Seasons
- Toys
- Water
- Celebrations
- Families
- Gardening/environment
- Homes
- People who help us
- Shapes
- Travel and transport
- Weather

The topic appears in a shaded box at the top of each page. The other books in this series use the same topics. However, these are only suggestions and all of the activities can easily be modified to fit with any topic that a practitioner needs material for. The majority of activities in this book can be used with any topic.

All activities are designed with the Statutory Framework for the EYFS in mind and, therefore, link with other learning and development areas, offering opportunities to explore them all with the children. Practitioners will be aware of needing to provide a balanced curriculum and to explore, along with Personal, Social and Emotional Development, Understanding the World and Mathematics, the areas of Communication and Language and Literacy, Physical Development and Expressive Arts and Design.

Prior knowledge is not expected for any of the activities, but practitioners should use their own judgement to choose activities to suit children's developmental stages.

Although plenary sessions have not been included, practitioners will recognize the importance of reviewing activities and encouraging children to verbalize what they did, how they felt about it and what they think they achieved. It is also important, of course, to discuss with parents and carers how they might build upon the children's experiences, so that they can learn consistently in the setting and at home.

Logos used on the activity sheets

Box 1 – group size

This box indicates the number of children recommended for the activity, keeping safety and level of difficulty in mind. Less able children can achieve more difficult tasks with a smaller child to adult ratio. The group size indicates the size of group for the activity itself, rather than for any introductory or plenary sessions.

Box 2 – level of difficulty

This box uses a scale between 1 and 5 to depict the level of difficulty or challenge the task might present to the children. Children still developing skills described in the 22–36 months age band of the Development Matters guidance document will find 1 most suitable, whilst 2 and 3 will apply to children as they move through the 30–50 months age band. Children just entering the 40-60+ months age band will appreciate 4, while 5 will be suitable for older and able children who are already meeting the Early Learning Goals. As most settings have mixed age groups, the majority of the activities have been classified as easy, so that the whole group can be involved. Higher levels can be achieved for particular children, as appropriate, by encouraging them to develop their own ideas and to participate in the suggested extension activities.

Box 3 – time needed to complete the activity

The suggested time slots are only a guideline. Children need time to practise their skills, test their ideas and reflect upon their findings. Some children will wish to extend the original activity to pursue their own enquiries or improve upon their experiment.

Safety

Where relevant, additional safety notes are included on the sheets. You are advised to read these before commencing the activity.

Safety

● Children are active learners, and investigative, exploratory and construction activities invariably involve the use of potentially dangerous equipment. Part of the learning process involves offering the child the opportunity to learn to use this equipment safely. As young children cannot anticipate danger, practitioners have to be vigilant and take part in a regular risk assessment exercise relevant to their own setting.

- Any rules issued by your employer or local authority should be adhered to in priority to the recommendations in this book; therefore check your employer's and local authority's Health and Safety guidelines and their policies on the use of equipment.

Links to home

- The word 'parent' is used to refer to all those persons responsible for the child, and include mothers, fathers, legal guardians and primary carers of children in public care. The 'Links to home' suggest ways in which parents can continue and reinforce the learning that is experienced at the setting.

- Parents can share important information about their children and their experiences, upon which practitioners can build. It is essential that practitioners find out from parents details of any special or additional needs, allergies, intolerances or medical conditions.

- Parents can be a valuable resource, giving support when extra help is needed during visits out of the setting, and with more complex activities during designing and making. They can become the knowledgeable visitor, bringing their own language, culture and experiences to the setting.

- Parents are also a useful source of recycled materials, which are required for many of the tasks.

Templates and other resources

On pages 190–204 there are photocopiable templates to be used in conjunction with relevant activities. The pieces will last longer if they are laminated.

Characteristics of effective learning can be found on pages 205–206.

The table of learning opportunities appears on pages 207–217.

An index of topics can be found on pages 218–220.

Assessment

- Each activity has learning objectives which are linked to the prime area of Personal, Social and Emotional Development and/or the specific area of Understanding the World and/or the specific area of Mathematics.

- To assist practitioners in planning a balanced educational programme of experiences, the charts on pages 207–217 show which activities address which of the Department of Education's EYFS Early Learning Goals.

- Comments on these activities and other evidence of children's achievements, such as dated examples of early writing, dictations, drawings, paintings and photographs of 3D models activities, experiments, investigations and 'work in progress', can be kept in a file or portfolio and given to parents as a celebration and record for the future.

- These records should be retained for inspection.

Personal, Social and Emotional Development

Self-confidence and self-awareness

- The prime area of Personal, Social and Emotional Development is vital to a child's well-being and later learning. Parents and carers must repeatedly and consistently offer the right experiences throughout the early years, in order to ensure healthy development of the brain, positive social interactions and constructive control of emotions.

- Babies are born without the ability to regulate emotions and need the full attention of caregivers who will respond quickly to soothe or remove any distress. This gradually creates brain connections and networks which allow children to believe that their needs will usually be met. This builds security and self-confidence, as they are reassured that they will be cared for and that their problems can be managed and handled.

- Children need to develop self-confidence so that they will feel comfortable with themselves and their abilities and positive about their learning. If they are willing always to attempt new challenges, to try hard, to do their best and to persevere when things are difficult, they will be able to believe in themselves and fulfil their potential.

- In order to achieve this, children instinctively know how to seek experiences that contribute to positive personal development. They stay close to parents and other familiar family members and caregivers, or key people in their early years settings, and demand the attention and interactions that they need to stay safe and to learn how to regulate emotions and relationships.

- Self-awareness involves understanding that love and respect are always available and can be relied upon, shared and offered to others. Children who are regularly and consistently shown love, affection, respect and empathy are stimulated to develop high levels of self-esteem, as they believe that they are worthy of love and regard for who they are. This positive attitude relies on their understanding that, while certain behaviours and abilities are preferred and others are undesirable, love is not dependent upon them, but is unconditional.

First visits

Topic
Families

Resources
■ A range of attractive toys and activities that a child could expect to experience in the setting
■ Information for parents.

Learning objectives
● To become familiar with the setting and their key worker in preparation for separation from their parent
● To express own preferences and interests
● To select and use activities and resources with support
● To develop confidence in asking adults for help

Links to home
● Invite children and parents to come to an introductory session for new children before they begin to attend the setting.
● Each child and parent can spend time with their key person, getting to know each other informally.
● A range of activities could be put out that both child and parent can do together.

● During the visits, the child can be shown where they will put their coat and what their special identifying symbol will be (take a copy home), and given an introduction to the routine that starts the session.
● The child can be shown the choice of activities available to them.
● Attending an open day, with children being involved in activities, will also help the child feel more familiar with the setting.

Related activities
● All the fun of the fair (see page 63)
● Sports day (see page 26)
● I'm here! (see page 23)

Personal, Social and Emotional Development with Understanding the World and Mathematics

Babysitter

Learning objectives
- To separate from main carer with support and encouragement from a familiar adult
- To describe self in positive terms
- To feel comfortable to talk about babysitters or other regular carers and about key adults in the setting
- To be confident to speak to others about own needs, wants, interests and opinions and to ask adults for help when needed

What to do
Circle time
- Ask children if they have a babysitter to look after them if their parents go out together on their own.
- Invite children to say what their babysitters are like.
- Read *Be Good Gordon* and ask whether anyone has a babysitter like Gordon's.
- Ask children whether they sometimes go to stay with grandparents.
- Read *Amy Said* and ask whether the children behave like that towards their grandmas.
- Discuss Amy's grandma's feelings about the children's bad behaviour and why they were good at the end.
- Ask children to think about whether their grandmas might sometimes complain about their behaviour.

Book corner
- Read *Small* and talk about what to take to grandma's home.
- Encourage the children to pack their bag for a weekend with their grandma.
- Talk about what they are taking.
- Read *That's My Mum*. Mia is of mixed heritage, and people think that her mum is her childminder.
- Encourage a child to talk about any problems relating to their mixed heritage.

Topics
Families/People who help us

Resources
- Books: *Be Good, Gordon* by Angela McAllister (Bloomsbury)
- Book: *Amy Said* by Martin Waddell (Walker Books)
- Book: *Small* by Clara Vulliamy (Picture Lions)
- Bags
- Dressing up clothes
- Small toys
- Book: *That's My Mum* by Henrietta Barkow (Mantra Lingua)

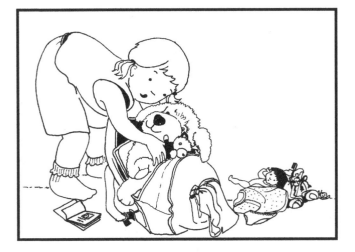

Links to home
- Determine the make up of families to make the activity inclusive.

Did you enjoy that?

Topic
Myself

Resources
- CD-ROM: '2count' from *2simple Infant Video Toolkit* (2Simple Software)
- Computer with colour printer and paper
- Pictures of cartoon characters and photographs of faces showing emotions
- Glue and glue spreaders
- Scissors
- Paper

⚠ Supervise the use of scissors.

Learning objectives
- To be confident to try new experiences and activities and to say why some activities are liked more than others
- To speak confidently to others about own opinions and how people show feelings

Preparation
- Do this activity after the children have encountered an activity for the first time.
- Select '2count' program.

What to do
Circle time
- Ask the children whether they enjoyed an activity.
- Talk about children's likes and dislikes.
- Talk about how people show their feelings through facial expressions and body language. Ask the children to show that they are happy, using their faces. Talk about their expressions.
- Discuss other emotions, inviting the children to think of as many different ones

as they can. Support the learning of new vocabulary and developing language skills by introducing words such as: *cross, angry, happy, sad, frightened, worried* and *surprised.*
- Explain that they are going to play a game where the other children have to guess their feelings.
- Invite a child to make a face of her choice showing an emotion.
- Encourage the class to guess the emotion.

Art activity
- Show pictures and photographs of people and cartoon characters demonstrating emotions.
- Ask the children to sort the pictures into those that are sad, and those that are happy, and make a collage.

Extension/variation
Computer activity
- Make a graph of feelings, showing 'Today I feel ...', using '2count'.

Links to home
- Ask parents to talk about feelings.
- Ask parents to donate unwanted comics and colour magazines.

Related activities
- Angry (see page 48)
- This is me (see page 80)

Colour matching

Learning objectives
- To show confidence in asking adults for help when needed
- To realize that a colour can have different shades and tones
- To take part in 'Show and tell'
- To become confident to speak within a familiar group about own ideas, interests and opinions

Preparation
- Collect objects and pictures to show to the children, to introduce different colours, shades and tones, and put them into a bag.
- Ask children and parents to create colour bags of their own – see Links to home.

What to do
Circle time
- Model for the children bringing items out of the bag one by one to talk about them.
- Over a period of a week, invite each child in turn to show the contents of their bag, and to point to and name the colours on each item.
- Ask them to identify the item and, if appropriate, say what it is used for or where they found it.
- Encourage the children to compare the shades and tones of the items. Support the learning of new vocabulary and developing language skills by introducing words such as: *darker, lighter, paler, brighter, greener, more yellow, navy blue, light pink* and *orange-red*.
- Talk about favourite colours.
Sort and match activity
- Sort the items on the collections table into groups showing the different colours.
- Alternatively, sort and match commercial toys (see Resources).

Extensions/variations
Book corner
- Read one of the books to the group.
Computer activity
- Make a graph of favourite colours.

Topic
Colours

Resources
- Collections table of items of different colours
- Toys: *Compare Bears, Fruity Fun Counters, Friendly Farm Animal Counters, Super Duper Sorting Set*
- Books: *Cat's Colours* by Jane Cabrera (Mammoth); *The Emperor Who Hated Yellow* by Jim Edmiston (Barefoot Books); *Chidi Only Likes Blue* by Ifeoma Onyefulu (Frances Lincoln)
- '2count' or '2graph' from *2simple Infant Video Toolkit* (2Simple Software) to make graph
- Computer with colour printer and paper

Links to home
- Ask parents to help their child to select three different items (not toys, and of little value) that have on them the same specified colour to bring into the setting in a labelled bag; for example, blue bottle top, blue sock, empty cereal box with blue writing. Ask different children to bring a different colour.
- Ask parents to reinforce the names of colours.

Self-confidence and self-awareness

What did you do at the weekend?

Topics
Families/Myself

Resources
- Chart showing days of the week
- Flashcards of the days of the week
- Books: *On Friday Something Funny Happened* by John Prater (Red Fox) or *Mr Wolf's Week* by Colin Hawkins (Mammoth)

Learning objectives
- To be aware that weekends with their family have a different routine from that during the week at the setting
- To talk freely about their home and community
- To recognize and describe special times and events for families and friends
- To understand similarities and differences between families, communities and traditions

What to do
Circle time
- Invite children to talk about what they did at the weekend. Offer support and prompts as necessary to help them to remember. Perhaps there is something they do every weekend, or perhaps each weekend is different? Ask whether any of them went to church or to a dance school, a swimming lesson or a sport or gym club, to visit friends or relatives, to a birthday party, or to the park to play.
- Repeat with the children the days of the week, using the words written on the flashcards.

- Support the learning of new vocabulary and developing language skills by chanting the days of the week in sequence and introducing words such as: *today, yesterday, tomorrow, last week, next week* and *weekend.*
- Read On *Friday Something Funny Happened* or *Mr Wolf's Week*.
- Encourage all the children to listen carefully as other children talk about their own experiences, cultures and faiths and model for them how to display interest and make positive comments.

Links to home
- Tell parents that the children will be telling the class on Monday what they did at the weekend. Ask them to remind their child what happened that weekend so that they will have something to contribute during circle time.
- Give each child a small booklet to record with mark-making or pictures, or notes written by the parents at the child's dictation, a diary of what they do at home each weekend.

Personal, Social and Emotional Development with Understanding the World and Mathematics

I'm here!

Learning objectives
● To separate from main carer, with support and encouragement and eventually independently and with confidence
● To have a sense of belonging at the setting
● To form positive relationships with adults and other children

What to do

Registration time
● Provide a coloured luggage tag labelled with each child's name on a table. Encourage and support each child to find her own tag and hang or stick it on the wall to show that she is present.
● To help with identifying the labels, add coloured shapes to them.
● Use the same coloured shapes to label the children's coat hooks and bags or folders.
● Remind children of the names of colours.
● Over time, remove the coloured shape.
● Praise those children who do not have the identifying coloured shape on their possessions, but recognize them by the name labels.

Circle time
● Discuss first days at nurseries, pre-schools, schools and other settings.
● Read *Sam's First Day*.
● Ask children if they can remember how they first found a friend at the setting. Read *You're All Animals*.

Mark-making activity
● Encourage the children to write their name at every opportunity.
● Encourage them to write their own name on the luggage tag.

Links to home
● Show each parent where their child's peg is situated.

Topics
Colours/Myself

Resources
■ Each child's coat peg labelled with coloured shape and their name
■ Book: *Sam's First Day* by David Mills (Mantra Lingua)
■ Book: *You're All Animals* by Nicholas Allan (Random House)
■ Named coloured luggage labels

Related activities
● Let me introduce myself (see page 61)
● My friend (see page 62)
● New friends (see page 39)
● First visits (see page 18)
● Hello (see page 65)

Honey cake

Topic
Food and shopping

Resources
- Book: *Eat Your Dinner!* by Virginia Miller (Walker Books)
- Song book and CD: 'Honey' from *Start with a Song* by Mavis de Mierre (Brilliant Publications)
- CD player
- Nursery rhymes: 'Pat-a-Cake' and 'Five Cherry Cakes in a Baker's Shop' from *This Little Puffin* compiled by Elizabeth Matterson (Puffin Books)
- Cakes: made during cooking activity
- Role-play area: set out as a kitchen with cooker and cooking implements and dining table with pretend food, crockery and cutlery

Learning objectives
- To be confident in new social situations and when trying new activities
- To be confident to talk to other children when playing and to ask adults for help when necessary
- To speak and express needs effectively, using good manners

What to do
Circle time
- Read *Eat Your Dinner!*
- Invite children to share experiences of not wanting to eat their dinner, or another meal, at home. Offer appropriate support to encourage all contributions and ensure that the children listen to each other.

- Ask whether the children like cake and which are their favourite types of cake.

Music activity
- Learn the song: 'Honey' and the nursery rhymes.
- Play the hand clapping game while singing 'Pat-a-Cake'.

Snack time
- Make sure that the children wash their hands first.
- Encourage them to be polite, and to say 'Please', 'Thank you' and ask before taking a cake.

Role-play activity
- Encourage role-play based upon the story.

Related activities
- Tea time (see page 59)
- At the café (see page 60)

**Personal, Social and Emotional Development
with Understanding the World and Mathematics**

I'm a builder

Learning objectives
- To be confident to try new activities, talk about own ideas and choose the resources needed for a particular project
- To understand behavioural expectations and boundaries, such as finding, using and putting away materials correctly and sharing resources with others
- To construct with a purpose in mind, selecting appropriate resources, tools and techniques to shape, assemble and join materials
- To learn positive attitudes and celebrate diversity through understanding that all careers are accessible to both genders

Topic
Homes

Resources
- Book: any *Bob the Builder* story
- Web site: www. bobthebuilder.org
- Resources as requested by the child

Preparation
- Ensure that resources are easily accessible to the children, and are stored so that they can be easily put away neatly.
- Encourage the children to find, use with care and return materials for themselves.
- Make space for the constructions to be stored so that the children can continue at a later time.

What to do

Circle time
- Read the *Bob the Builder* story.
- Talk about how more women are becoming builders.
- There could be a discussion about popular television programmes dealing with improving homes and gardens.

Design and making activity
- Encourage the children to design and make houses, gardens and garages from any resources that they choose.
- Support them as they collect the resources themselves.
- Discuss the different parts of the house and the shapes and positions of the roof, doors and windows.
- Encourage her to store her construction safely, so she can continue building it during a later session.
- Take a photographic record of the construction.

- Encourage the children to put away unused resources. When the buildings are finished, they should dismantle those made from construction kits and put the pieces away neatly. Models made from junk and craft pieces could be taken home.

Related activities
- Straw shapes (see page 30)
- I want to be (see page 27)

Sports day

Topic
Gardening

Resources
- Beanbags
- Tunnels, tables or cardboard boxes
- Per team: small buckets and 2 larger buckets (one with water, one to fill)
- Sand
- Prizes
- Certificates

⚠ Do a risk assessment. Lay out the grounds so that spectators do not obstruct children. Check first-aid kit. Check chronological ages for physical development.

Learning objectives
- To develop a positive self-image and describe self and talk about own abilities confidently while taking part in a group event
- To be confident to try new activities and talk about liking some more than others
- To move confidently in a range of ways, showing good control and coordination and safely negotiating space within a group

Preparation
- Organize the timetable for the event.
- Buy prizes and print certificates (enough for all the children)

What to do
Circle time
- Consult the children and choose a theme, in this case 'In the Garden'.
- Involve the children by asking for ideas for games.

Outside physical activity
- Practise the games for the day.
- Where appropriate, choose cooperative team games. Some examples might be:
  - Scattering seed: how far can they throw a beanbag?
  - Creepy-crawly race: children to travel on their hands and knees.
  - Worms (obstacle course of tunnels): climb through tunnels, collect leaf, and climb back through tunnels. If no plastic tunnels, use large cardboard boxes, or tables in a line.
  - Water the garden: run with bucket of water, and fill a larger container.
  - Frog jump: children have to jump with both feet together on to sand pit.
  - Wheelbarrow race: two children, one as the wheel barrow with their legs held by the other as they 'walk' along on their hands.

Links to home
- Send leaflet to parents with time and date, and alternative dates if there is rain.
- Remind parents to supply sun cream and hats if it is hot on the day.

Related activities
- All the fun of the fair (see page 63)
- First visits (see page 18)

I want to be ...

Learning objectives
- To be confident to try new activities and express own preferences and interests
- To develop a positive self-image and talk about own abilities and opinions
- To learn positive attitudes and celebrate diversity through understanding that all careers are accessible to both genders
- To show interest in different occupations and ways of life

Preparation
- Share books and pictures with the children showing images of both men and women in a variety of different careers.

What to do

Circle time
- Read *ABC I Can Be* and show pictures/ photographs. Ask what job the people are doing, and where.
- Ask what job the children would like to do when they grow up. Ask for reasons.

Table activity
- Play 'Happy Families' with small groups of children, emphasizing that all members of each family, of both genders, follow the same careers.

Role-playing
- Encourage the children to dress up and say which job they are representing.
- Ensure that boys feel comfortable to choose to play at being nurses, dancers and chefs and that girls may happily choose to dress up as fire fighters, builders and racing drivers, etc.
- Encourage both boys and girls to construct and play with diggers in the sandpit and to bath dolls in the water tray and put them to bed in the role-play area.

Topic
People who help us

Resources
- Story books: *ABC I Can Be* by Verna Allette Wilkins & Zoë Gorham (Tamarind Books)
- Non-fiction picture books and pictures showing places and people working
- 'Happy Families' card game
- Dressing-up clothes and hats (careers)
- Sand play, building site toy vehicles
- Book: *Dig Dig Digging* by Margaret Mayo and Alex Ayliffe (Orchard)
- Construction kit to make vehicles
- Water tray: washable dolls, flannels, towels, soap

 Check local authority or setting guidelines for 'out of setting' visits.

Links to home
- Ask if any parents work in or have connections with a fire station, a building site, a hospital or another suitable place and could possibly arrange for the children to make a safe visit.

Related activities
- I'm a builder (see page 25)
- Thank you (see page 71)

Stranger danger

Topic
Health

Resources
- Large safe space
- Web site:
  www.kidscape.org.uk
- DVD's available from
  Kidscape: *Cosmo and
  Dibs Keeping Safe*
  (five short, fun safety
  stories for ages 3–6 with
  teaching notes); *Be Free,
  Be Safe, Be Smart*

Learning objectives
- To develop self-confidence and the ability to understand differences between acceptable risk and possible danger
- To understand the need for safety and practise some appropriate safety measures without direct supervision

What to do

Circle time
- When appropriate, talk to the children about 'stranger danger'.
- Discuss the following safety rules with the children:
  - Never be away from your home on your own.
  - Never leave with anyone (even if you know them) without your parents' permission.
  - Don't talk to strangers when you are not standing with a parent, because they might persuade you to go with them. The bad people who do this might offer to give you sweets in their car, or ask you to help them find a puppy.

- If you feel that someone might harm you, shout and scream very loudly.
- Run and try to get away. Attract someone's attention.
- Try to remember what they look like, and always tell your parents. Parents need to know about problems to help you.
- Some secrets should never be kept. No-one, even someone you know, should ever ask you to keep a kiss, hug or touch of any kind a secret. Always say no if anyone tries to touch you in a way that frightens or confuses you.

Role-playing
- Practise saying 'No!' and running to safety.
- Where would you go where it was safe?
- Ask children to think of places that they could go to in order to be safe.

Links to home
- Tell parents that you are talking about 'stranger danger' so that they can reinforce the warning.

Circle time

Learning objectives
● To develop self-esteem and self-confidence
● To speak in a familiar group, talk about own ideas and abilities and describe self in positive terms
● To have a developing awareness of their own needs, views and feelings, and sensitivity to the needs, views and feelings of others
● To form positive relationships with adults and other children

Topic
All topics

Resources
■ Soft toy, puppet, toy microphone to pass from one child to the next to help them speak to the group
■ Article(s) under exploration to help make learning concrete for the children
■ Story book to introduce or support topic
■ Sufficient space for the children to sit on the floor in a circle
■ Chair for practitioner
■ Flip chart and marker pen if required
■ Small table to place articles

What to do

Circle time
● Circle time can improve self-esteem and hence self-confidence through celebrating achievements and sharing information.
● Give praise and encouragement, pointing out why it is worthy of praise. Being valued and loved boosts positive self-esteem.
● Looking, listening, speaking in turn and joining in are central to the ethos of circle time.
● Maintain eye contact while speaking or listening to a child. Ensure that all the children are listening to other children's comments. Allow children to 'pass' if they don't wish to say anything, but give plenty of support to timid children. Use a puppet, soft toy or toy microphone to pass from child to child round the circle to encourage children to speak one at a time.
● The physical closeness and working together can enhance feelings of belonging to the group.
● The circle shape ensures that everyone sees everyone else. Encourage everyone to be involved.
● Circle time can improve thinking skills as well as spoken language.
   ◆ Encourage the children to 'have a guess' and praise their ability to learn from their mistakes.
● By sharing problems, the children can develop empathy.
   ◆ Encourage them to think through problems. Support them in their attempts to orally express the solution to their problem.

● By having the opportunity to compare experiences and views, the children can become aware of their own needs, views and feelings as well as those of others.
   ◆ Help children to understand why there are similarities and differences between their experiences and points of view.
● Encourage the children not only to answer the practitioner's questions, but also to ask questions of the practitioner and of the other children.
   ◆ A child needs to know part of the answer to be able to formulate the question, so needs reassurance and to receive a proper response to their questions. If you don't know the answer, say so, and suggest that you both try to find the answer.

Related activity
● Most of the activities in this book begin with a circle time.

**Personal, Social and Emotional Development
with Understanding the World and Mathematics**

Straw shapes

Topic
Shapes

Resources
- Book: *Three Little Pigs* by Nick Sharratt and Stephen Tucker (Macmillan), a lift-the-flap book, or *The Three Little Pigs* adapted by Elizabeth Laird (Heinemann Library)
- Book: *Shapes* (Dorling Kindersley) – lift-the-flap book
- Photographs of bridges with steel struts and diagonal supports
- Construction toy: *Fun Straws*

Learning objectives
- To talk about own ideas and choose the resources needed for particular activities
- To speak within a familiar group and ask for help when needed
- To use and explore a variety of materials, tools and techniques and to experiment with design, form and function

What to do
Circle time
- Read *Three Little Pigs*.
- Talk about using different materials to make a house.

Construction activity
- Point out the different shapes the children have made. Together, determine if the shapes are strong or whether they are weak and will collapse.

- Note whether the children are persistent, and whether they enlist help from their peers and/or adults to achieve their aim.
- Encourage the children to explore and talk about their ideas and findings, rather than what they are making. Accept their ways of solving problems.
- Praise their efforts and invite them to 'show and tell' the rest of the class.
- Allow space to store their construction so that they can return to it and complete the task during a later session.

Extensions/variations
- Ask what else could be made using the construction straws.
- If appropriate, talk about how bridges have to be light, but very strong.
- Show pictures of bridges, pointing out the steel structures, and especially the diagonal struts.
- Encourage the children to make a bridge using straws.

Related activities
- I'm a builder (see page 25)

Personal, Social and Emotional Development

Managing Feelings and Behaviour

- During the first five years of life, children grow from babies who are completely unable to control or tolerate any form of distress to young people who manage and contain their own feelings, emotions and behaviour on a day-to-day basis.

- Caregivers need to be consistently available to children to help them to contain their emotions and reassure them that their feelings are not too big or dangerous to be controlled and handled. For brains to develop strong connections, allowing children to control their own emotions, they must have good, positive experiences to build on. If adults provide an enabling environment, model appropriate behaviour and ensure that children experience calm and loving interactions and empathy and respect for others, the children will be able to develop these skills and dispositions for themselves.

- Positive experiences stimulate the production of hormones that make the children feel good and therefore encourage them to repeat the experiences, while unpleasant experiences and emotions, such as tantrums or friends being upset, may activate stress hormones and therefore discourage repetition.

- Building a secure relationship with one or two key people is essential for children and families who are attending an early years setting. It is important that all primary and secondary key caregivers know and understand their children well enough to recall and link experiences and to recognize the causes of their various reactions and behaviours.

- Acknowledging and talking about feelings teaches children that they are powerful but also acceptable and ever-changing. Sensitive adults are able to distract, reassure or simply comfort children in distress, helping them to move away from a feeling that life is unbearable, through an understanding of why they feel sad or angry, and onto thinking of a reasonable solution, such as finding another friend or activity, sharing or waiting for a turn.

- This is more effective than labelling behaviour driven by emotion as 'good' or 'bad', or expecting young children to control their brains' reactions to what they perceive as threatening before they are physically able to do so.

The tree house

Topic
Homes

Resources
- Book: *Nearly but not Quite* by Paul Rogers (Red Fox)
- Role-play area
- Outdoor climbing frame

⚠ Supervise the use of climbing frame.

Learning objectives
- To accept the needs of others and respond to their feelings and wishes
- To understand that actions affect others and begin to give help, support and comfort to other children when they need it
- To adjust behaviour to fit different situations
- To represent ideas, thoughts and feelings through role-play and stories

What to do
Circle time
- Read *Nearly but not Quite*.
- Ask the children whether they would like to play in a tree house, like Simon.
- Discuss the story with the group, asking questions such as:
  - ◆ Did Simon keep up with the older children? (He tried to. They did help him sometimes.)
  - ◆ What happened to Simon? (He couldn't keep up with James and Harriet because he was not as big. He did not know how to ask for help. He got very dirty and wet.)
  - ◆ What could he have said to them to make them understand that he could not keep up with them?
  - ◆ Did Simon enjoy his time with James and Harriet? (He wanted his mummy.)
- Ask whether the children would try to keep up with the older children, or give up?
- Ask if the children have felt upset when they could not do something because it was too difficult for them.
- Talk about how difficult it is for some children to come to the setting and leave their mummy. Ask the children what they could do to help such a child.

Imaginative play
- Encourage a mixed-age group of children to be involved in imaginative play using the role-play area and/or a climbing frame.
- Encourage the children to give support to each other and the older children to help the younger ones, if they need it.
- Ask what their game is about. The practitioner should join in to expand and extend their play.
- Support the learning of new vocabulary and developing language skills by introducing words and phrases such as: *feelings, support, help me please, wait for me, slow down, make room for me please* and *when may I have a turn?*

Links to home
- Discuss with parents how their child is settling into the setting.
- Negotiate with parents how best to make their child feel secure on being left at the setting.
- If a child finds it difficult to separate from his/her parent, suggest that the parent says 'See you soon. Have a nice time,' rather than 'Goodbye'.

Related activities
- Babysitter (see page 19)
- First visits (see page 18)

Personal, Social and Emotional Development with Understanding the World and Mathematics

Underground

Learning objectives
- To talk about feelings and behaviours and understand how actions affect other people
- To adjust behaviour to suit different situations
- To confidently try a new activity and choose resources to extend the play
- To engage in imaginative role-play, with or alongside other children

What to do
Circle time
- Read *The Picnic*. Initiate discussion with the children and ask questions to encourage exploration of the topic. For example:
  - Which animals live underground?
  - Are you afraid of the dark?
  - How do you feel when it is dark?
  - Does any dark place frighten you?
  - Describe your dark place that frightens you.
  - Is there any dark place that does not frighten you?
- Discuss how dark and confined it must be underground, but the animals feel safe and secure there because it is their home.

Making relationships
Imaginative play activity
- Encourage the children to use the large cardboard boxes and use blankets to cover tables to make tunnels through which they can play.
- Suggest that the children imagine themselves as animals that live underground.

Topic
Animals

Resources
- Book: *The Picnic* by Ruth Brown (Andersen Press) shows animals that live underground
- Non-fiction books about animals that live underground
- Large cardboard boxes
- Blankets to cover tables
- Collect real worms (return to garden when activity is finished)
- Magnifying glass
- Pasta, cotton reels, coloured beads
- Playdough
- String
- Dice with numbers

 Wash hands after examining the worms.

Extensions/variations
Craft activity
- After examining real worms, using a magnifying glass, make worms by threading pasta, coloured beads or cotton reels on to string.
- This activity can be turned into a game if children take turns to roll a die and add that number of beads to their own strings.
- Playdough can be rolled into a thin sausage to make a worm.

Personal, Social and Emotional Development with Understanding the World and Mathematics

Allsorts

Topic
Colours/Shapes

Resources
- Liquorice allsort sweets, for all the children
- One example of each type of liquorice allsort sweet
- Playdough coloured pink, black, yellow, blue, orange
- Rolling pin
- Square cutters, knife
- White poster paint
- Paintbrushes
- PVA glue
- Small plates
- Paper bags

⚠ Supervise cutting tools. Check children's records for allergies and diabetes.

Learning objectives
- To be aware of behavioural expectations and boundaries
- To adjust behaviour to different social situations and conventions
- To work cooperatively with a partner or small group to create models
- To show sensitivity to the ideas and feelings of others and take account of what they say while organizing an activity
- To handle tools and malleable materials with increasing control

What to do
Circle time
- Ask the children to wash their hands.
- Tell them that they will need to think how they are expected to behave during the next activity.

- Pass round some liquorice allsort sweets on a plate, without saying anything. (No-one should take a sweet.)
- If any child does take a sweet, gently ask the group whether the practitioner had said that they could.
- Then tell the group that they are investigating the sweets in preparation for an art activity.
- Ask them what shapes they are, their smell, and finally, when everyone has looked at the sweets, what they taste like.
- Ask the children whether they think knowing about the taste will help them to make better models.
- Accept all ideas.
- Support the learning of new vocabulary and developing language skills by introducing, for example: colours – *pink, yellow, blue, black, white;* shapes – *rectangle, square, cylinder, sphere (hundred and thousands);* sizes – *bigger than, smaller than, the same, about the same;* textures –*Allsorts smooth, rough, bumpy;* layers – *one, two, three, sandwich;* smell – *Turkish delight, rose, coconut, liquorice.*

Art activity
- Arrange some liquorice allsort sweets on a plate.
- Ask the children to work cooperatively with friends to make playdough sweets.
- Encourage them to use the liquorice allsorts as an example, but reassure them that their models don't have to be exactly the same size as the real sweets.
- Suggest they divide the task of making the sweets between them.
- Make playdough sweets and bake.
- Paint the white layers with white poster paint mixed with a little PVA glue.
- Arrange the sweets on a plate or in a paper bag.

Related activities
- Can I have a biscuit, please? (see page 35)

Can I have a biscuit, please?

Learning objectives
● To accept the needs of others and share resources
● To adapt behaviour to suit different events and social situations
● To understand that actions affect other people
● To begin to negotiate and solve problems

Preparation
● Produce a notice to tell parents about a 'sharing picnic', asking them to supply food and drink.

What to do
Circle time
● Ask the children to wash their hands before coming to the circle.
● Tell them that they have to listen very carefully to your instructions.
● Ask the children to sit together, facing each other in pairs.
● Place a napkin between each pair, and place on it one biscuit.
● Tell the children that they cannot take the biscuit for themselves. Ask for suggestions of ways to share the biscuit.
● Then put out a second biscuit.
● Talk about their decisions.
● Ask whether the children like sharing with friends, whether it feels easier to share with some people than others and what happens if they don't share.

Extensions/variations
Sharing picnic
● Suggest that the children have a 'sharing picnic', where everyone brings enough food for two children, to share with the class.
● The children can 'show and tell' what they have brought, and explain the reasons for their choice.

Topic
Food and shopping

Resources
■ Biscuits for all of the children
■ Napkins

 Check children's records for allergies and diabetes.

Links to home
● Each child should go shopping with their parent to choose some food and drink. Ask parents to involve their child closely in choosing their contribution to the picnic.
● Ask parents to discuss with their child the reasons for their choices, eg whether the food is their favourite and they wish to share it, or whether they are choosing something special that they think the other children might enjoy.

Related activities
● Allsorts (see page 34)

Stepping stones

Topic
Gardening

Resources
- Grassy area with stepping stones of different materials set into the grass
- Blindfolds
- Book for practitioner: *Being Blind* (Chrysalis)

Learning objectives
- To accept the needs of others and adapt behaviour to different situations
- To understand how own actions can affect other people and begin to offer help and support to others when appropriate
- To know about similarities and differences between people and why some people may need extra help, support or understanding in particular situations

Preparation
- The secret garden should have stepping stones of different materials set into the grass.

What to do
Role-playing
- This can be done outside in the garden.
- Pair off the children.
- Invite children to choose partners, with support, and prepare for an activity carried out in pairs.
- One should close their eyes or, if they can, have a blindfold, and have bare feet.
- The other child will help them safely walk over the stepping stones by holding their hand, and making sure that they don't stub their toes.
- Afterwards, talk about what sensations they felt.

- Ask whether they used other senses more while they couldn't see. For example, did they listen very carefully?
- Support the learning of new vocabulary and developing language skills by introducing words such as: *scratchy, crunchy, rough, smooth, cold, warm, tickly, scared, frightened, alone* and *dark*.
- Encourage children to guess what materials the stepping stones are made from, if they are not all concrete, and to say how they could tell by touching them.
- Ask the children who were guides what it was like supporting the 'blind' children and how difficult it was to help them to walk safely.
- Ask whether the children have any family members or friends who are blind or partially sighted and encourage them to discuss and describe their own experiences. (Be very sensitive if any partially sighted children attend the setting, but especially encourage them to talk about their own experiences and feelings and how and when they would like people to help them.)
- Talk about what it must be like to be blind.
- Retell *Being Blind*.

Related activities
- Making the secret garden (see page 66)
- Underground (see page 33)
- Keep the noise down (see page 47)

Laughing

Learning objectives
● To begin to accept the needs of others, with support
● To inhibit own actions or reactions and adjust behaviour to suit different situations
● To show sensitivity to others' needs and feelings and form positive relationships

What to do

Circle time
● Read the poem 'Laughing Time'.
● Ask the children if they think animals laugh.
● Ask children to think of things that make them laugh. Write down what they say and make a list.
● Read *Little Oops!* and discuss the difference between laughing at the antics of Preston Pig and the Wolf, and laughing at a person falling over or being tripped up on purpose.
● Sometimes people should not laugh. Ask the children when it might be inappropriate to laugh, eg when someone has fallen over and is hurt.

Art activity
● Encourage the children to paint a picture showing something that makes them laugh.
● Talk to them about their picture while they are painting.
● Ask them why it makes them laugh. Explore with them whether it was a kind thing to do. Were they laughing at someone, or laughing with them?

Topic
Myself

Resources
■ Poem: 'Laughing Time' by William Jay Smith from *Noisy Poems* illustrated by Debi Gliori (Walker Books)
■ Book: *Little Oops!* by Colin McNaughton (Picture Lions)
■ White board or flip chart and marker pen
■ Paper, paints and paintbrushes

Well done

Topic
Colours

Resources
■ Items that are made from gold, silver and bronze
■ Medals (variety of types)
■ Certificates of merit

Learning objectives
● To talk about the feelings and behaviour of self and others and their consequences
● To work as part of a group, understanding and following rules and expectations adjusting behaviour to different situations
● To welcome and value praise for behaviour and achievements
● To enjoy the responsibility of carrying out specific tasks
● To describe self and others in positive terms and talk about abilities, demonstrating a sense of pride in own achievements

Notes for practitioner
● Stickers, stars and certificates of achievement are a visable sign that effort and achievement are valued.
● Anything positive can be rewarded immediately with a sticker.
● Children prefer to display stickers on their chest as a badge of achievement, so make sure that they remain stuck on, to prevent tears later!
● There should be no comparison between children to make the acquisition of stickers a competition.

● An accumulation of stickers could indicate a child's motivation to please, so ensure that your response is genuine, and the child's work or behaviour is of worth. There should be evidence to support the praise; for example, 'You helped to clean the work table', rather than, 'You have been helpful today'.
● Take note if the child responds to your praise.

What to do
Circle time
● Talk about precious metals, and why they are expensive to buy.
● Show metals – gold, silver, bronze – and comment on their colours.
● Talk about medals that are given to people. What have these people done to win the medals? (Something special.)
● Explain that the children can gain medals, or badges, for doing good work, or behaving exceptionally well, such as being helpful to the members of the group.
● Show samples of the kinds of stickers that they can be given.
● Make a certificate of merit to give once a week to older children.

Related activities
● How thoughtful (see page 69)
● Be kind (see page 70)
● Thank you (see page 71)

New friends

Learning objectives
● To learn to adapt behaviour and adjust feelings to different social situations and to accept changes within routines and relationships
● To take steps to resolve conflicts with other children
● To show sensitivity to others' needs and feelings and form positive relationships

Preparation
● Use this activity after the practitioner has observed a dispute between children because a third child is now a new friend of one of them.

What to do
Circle time
● Read *Ebb's New Friend* and then initiate a discussion with the children, asking the following questions:
  ◆ How did Ebb the dog feel about Bird? (She felt jealous.) Give reasons for your answer. (Ebb thought that Flo wasn't her friend anymore, now Flo had Bird.)
  ◆ How did you feel when your best friend got a new friend?
  ◆ Why can't you be friends with both of them?
  ◆ When Bird flew away, how did Ebb feel? (Ebb missed Bird.)
  ◆ What happened when Bird came back? (They shared the favourite place at the front of the boat.)
  ◆ Were Flo, Ebb and Bird friends at the end of the story?

Role-playing with puppets
● Initiate an imaginative role-play with the children who have disagreed or been upset, using the puppets, based on the real-life situation. A practitioner could take on one of the roles and support the children in the others.

Topic
People who help us

Resources
■ 'Book: *Ebb's New Friend* by Jane Simmons (Orchard Books)
■ Hand puppets

● Encourage the children to respond in character and to discuss why they stopped being friends and to consider all being friends together.
● Invite the children to show their puppet play to the rest of the group.

Related activities
● Let me introduce myself (see page 61)
● My friend (see page 62)
● Hello (see page 65)

Not everyone likes fireworks

Topic
Celebrations

Resources
- Resources pack: People for the Ethical Treatment of Animals Research and Education Foundation
- Book: *Flash, Bang, Whee!* by Karen Clark, illustrated by Ian White (Mantra Lingua)
- CD-ROM: '2publish' from *2simple Infant Video Toolkit* (2Simple Software)
- Computer with colour printer and paper
- Paints, paintbrushes and paper

Learning objectives
- To understand that behaviour and actions have consequences and affect others
- To think about issues from the viewpoint of others
- To know that not all children enjoy the same things and to be sensitive to this
- To show care and concern for others, living things and the environment

Preparation
- Select '2publish' from *2simple Infant Video Toolkit*.

What to do
Circle time
- This activity is suitable for early November. Initiate a discussion with the children by asking:
  - Who is having a 5ᵗʰ November/Guy Fawkes/Bonfire Night party?
  - Are you having a bonfire with fireworks?
  - Do you like fireworks? Which type are your favourite?
  - Does everyone like fireworks?
  - What should you do with your dog or cat on Bonfire Night?
- Ask children to suggest any naughty or silly things that people might do with fireworks and to say why they are very dangerous and what could happen. Make sure that they all understand.
- Suggest to the children that you could make some posters together, to tell people what could happen and help them to be safe/kind on Bonfire Night?
- Encourage the children to make suggestions.
- Read *Flash, Bang, Whee!*

Computer activity
- Make a drawing for a poster.
- Support typing in the text.

Role-playing
- Observe whether the children can differentiate between safe and dangerous activities.

Art activity
- Encourage the children to paint a picture of their Bonfire Night party.

Stop, look, listen

Learning objectives
● To learn the importance of being safe near roads and how to achieve this
● To understand behaviour and its consequences and that some types of behaviour are unacceptable and unsafe
● To work as part of a group to follow rules and adjust behaviour to fit different situations
● To engage in imaginative small-world and role-play based on first-hand experiences

What to do

Circle time
● Show pictures of roads with cars and lorries. Ask the children what they would see on roads and pavements, what they would hear and what they might smell.
● Talk about the Green Cross Code, and stress using their senses. Talk about the sound of the signal to cross, and the colours of the lights.
● Ask who might use the sense of touch to help them cross the road safely? (Blind person could feel the bumps on the pavement to tell that there is a pelican crossing.)
● Ensure that children can tell the difference between the road and the pavement.
● Ask children if they know why they shouldn't stand right at the edge of the pavement? (Sometimes vehicles do come on to the pavement when they turn a tight corner.)
● Ask what rules they should follow to be safe and not get knocked down by a vehicle, for example:
    ◆ Only go out with an adult or someone who is old enough

Topic
Transport and travel

Resources
■ Book for practitioner: *Stop Look Listen Live* (Department for Transport)
■ Worksheets and pamphlets from www. hedgehogs.gov.uk
■ Pictures of roads with cars and lorries
■ Play mat showing houses and roads
■ Toy cars and small-world people

    ◆ Play and ride bikes in the garden or the park, never on the pavement or on the road
    ◆ To cross the road, stop at the kerb, look and listen, then walk carefully, holding an adult's hand.

Imaginative play
● Play on the play mat with cars and small-world people.
● Note whether the play people are walking along the pavement, and crossing the road with care.

Links to home
● Ask parents to reinforce rules for being safe near the road.

Potato latkes

Topic
Celebrations/Food and shopping

Resources
- Potato peeler
- Grater
- Knife or vegetable chopper
- Chopping board
- Tablespoon
- Frying pan
- Kitchen paper
- Plate
- Ingredients for potato latkes:
  6 large potatoes
  1 medium onion, chopped
  2 beaten eggs
  1 tsp salt
  1/2 tsp pepper
  1/2 cup flour (matzo meal if possible)
  Cooking oil for frying
  Apple sauce

 Check children's records for allergies and diabetes.

Learning objectives
- To adjust behaviour to different situations and understand and follow rules of safety and good manners
- To work as part of a group, taking turns and sharing resources
- To share and celebrate special practices to form positive impressions of cultures and faiths within the community
- To appreciate the need for good hygiene during food preparation and meal times

What to do
Cooking activity
- Wash your hands before you begin.

To make potato latkes
- Peel potatoes and soak in cold water.
- Grate potatoes and remove excess liquid.
- Supervise adding salt.
- Mix in other ingredients.
- Fry small spoonfuls until golden brown.
- Drain and serve hot with apple sauce.

Snack time
- Encourage the children to lay tables, serve and clear away.
- When handing round the food, encourage them to say 'Please' and 'Thank you'.

Links to home
- Ask for help with the cooking activity.

Don't do that

Learning objectives

● To talk about behaviour and its consequences and know that some behaviour is unacceptable

● To understand and follow rules as part of a group

● To be aware that they need to care for themselves to stay healthy

What to do

Circle time

● Read *Don't Do That, Kitty Kilroy* and then initiate a discussion with the children, asking the following questions:

◆ How did Kitty speak to her mummy?

◆ How do you think her mummy felt?

◆ Do you think that Kitty's mummy wanted to spoil her fun?

◆ What did Kitty do?

◆ How did Kitty feel afterwards?

◆ Is it a good idea to eat what you want?

◆ Is it a good idea not to go to bed?

◆ Is it a good idea to behave as you want?

◆ Do you think that Kitty's mummy was really letting Kitty do as she wanted?

◆ How should you speak to your mummy and daddy?

◆ What do you think that you should eat? Why?

◆ How much sleep should you have? Why?

◆ How do you think you should behave at home?

◆ Should you wait for your mummy to tell you not to do something? Why?

◆ How should you behave at the setting?

◆ Can we write down some rules on how to behave?

Topic
Health

Resources
■ 'Book: *Don't Do That, Kitty Kilroy* by Cressida Cowell (Hodder Children's Books)
■ White board or flip chart and marker pen

● Praise positive responses.

● Agree on drawings to represent the rules, such as 'Help each other', 'Keep things tidy', 'Share' and 'Listen when people are speaking'. Ensure that they are positive rules – not beginning with 'Don't'.

● Display the rules prominently.

● Children need guidelines on how to behave appropriately, showing what should be done rather than just a negative 'Don't do that'.

● Attract their attention by using their name, standing by them rather than calling across the room.

Role-playing

● Observe whether the children incorporate the story into their play.

Related activity

● The four Cs (see page 56)

It's mine!

Topic
Toys

Resources
- Two glove puppets (Tick and Tock)
- Basket
- One radio-controlled car
- Book: Good Friends series: *It's Mine* by Janine Amos (Cherrytree)
- Sand timer 5 minutes

Learning objectives
- To begin to be able to negotiate to solve problems
- To understand how actions affect other people

What to do
Circle time
- Using the two puppets, Tick and Tock, role-play the scenario that Tick has 'borrowed' Tock's car.
- Practitioners could speak for the toys and also directly to the children, using the following ideas:
  - 'I can't find my car anywhere. Have you seen it children?'
  - Tock is very sad and frustrated because he can't find his car.
  - How would you feel if someone had borrowed your toy?
  - What should you do if you want to borrow a toy?
- Then Tock finds his car in Tick's basket:
  - 'That's my car!'
  - 'But you weren't playing with it,' says Tick. 'I'm playing with it now.'
  - 'It's my car, give it back to me,' says Tock. (Tug-of-war with the car.)
  - 'Is it really your car?' asks Tick.
  - 'Well, no. It belongs to the setting,' says Tock.
- What should Tick and Tock do?
  - 'We will have to share the car. It's your turn.'
- How will they decide how long Tock can play with the car?

Outdoor activity
- Allow three children to share playing with the remote-controlled car.
- Observe how they are successful in sharing the toy.

Book corner
- Support reading *It's Mine*.

Related activity
- Angry (see page 48)

Me first

Learning objectives
● To be aware of behavioural expectations and to understand and follow social rules, such as making a queue or waiting your turn
● To form positive relationships with other children

Preparation
● Load the CD-ROM *All About Number Level 1*, or any other suitable game that includes ordering numbers.

What to do

Snack time
● Before snack time, or at another suitable moment, ask the children to line up in a single queue.
● Count the children in the line.
● Ask who is first/last in the line.
● Support the learning of new vocabulary and developing language skills by introducing the sequence of ordinal numbers: *first, last, second, third,* etc.
● Talk about why it is important to line up and take turns.

Computer activity
● Allow no more than four children to play on the computer.
● Ask who is going to use the computer first/second/third/fourth.
● Check that the children are taking turns.

Role-playing
● While playing shops in the role-play area, encourage the children to line up at the counter, and talk about their position in the queue.

Table activity
● Encourage the children to play fairly at the *Supermarket* game.
● Encourage those children who know how to play the game to support those who have not played it before.
● Praise positive behaviour.

Topic
Food and shopping

Resources
■ CD-ROM: *All About Number Level 1* (SEMERC)
■ Role-play area: set up as a shop
■ Game: *Supermarket* by Pam Adams (Child's Play)
■ Cash register and money

Related activities
● At the café (see page 60)
● How thoughtful (see page 69)
● Be kind (see page 70)
● Thank you (see page 71)
● The four Cs (see page 56)
● Hello (see page 65)

Nightmares

Topic
Health

Resources
- Book: *Where the Wild Things Are* by Maurice Sendak (Red Fox)
- Book: *Knock, Knock Who's There?* by Sally Grindley, illustrated by Anthony Browne (Puffin)
- Dream catcher
- To make a dream catcher: wooden curtain rings (make small notches to catch the wool), woollen thread, plastic bodkin, cotton thread, beads, small feathers
- Scissors

⚠ Supervise use of scissors and bodkins.

Learning objective
- To express own feelings and talk about how people show feelings

What to do

Circle time
- Read the book *Where the Wild Things Are*.
- Initiate discussion with the children, asking the following questions:
  - The naughty boy is sent to bed. Does he have nightmares? In the story the boy becomes the monsters' king.
  - The boy goes back home – and wakes up with his supper still hot. Was it a dream? Or a nightmare?
- Do you have nightmares? Tell us about them. Did your heart pound? Did you shake?

- Have you wondered who was at your bedroom door?
- Read *Knock, Knock Who's There?*
- It was Daddy at the bedroom door!
- Show the children a dream catcher and tell them that North American Indians used to hang them by their beds and believe that they would catch all the nasty nightmares

Design and making activity
- Make your own dream catcher (see diagram).
- Support tying the ends.
- Decorate with beads and feathers.

Personal, Social and Emotional Development with Understanding the World and Mathematics

Keep the noise down!

Learning objectives
- To be aware of boundaries set and behavioural expectations, such as when and where to be noisy or quiet, understanding that actions have consequences and affect other people and some behaviour is unacceptable
- To feel empathy for the deaf and those with communication difficulties

What to do
Circle time
- Discuss noise and quiet with the children and ask them if they like noise. Invite them to shout 'Yes' if they do and whisper 'No' if they don't.
- Talk about places where loud noises are acceptable, such as football matches, or if a child thinks she is in danger and needs to shout 'No!' as loudly as she can.
- Read about noisy things and then ask children the following questions:
  - Are you noisy at home?
  - Why should we play quietly inside the classroom?
  - Why should you wait before someone has stopped talking before you speak?
  - How can you mean 'Yes' and 'No' without speaking?
- Agree on a silent signal that the practitioner can make, so that the children know to be silent.
- Encourage the children to respond quietly in telling the children who have not noticed the signal to be quiet.
- Practise the signal to be quiet. Give praise.
- Now invite children to close their eyes and ask them: What can you hear? Can you hear your heart beat? Can you hear the birds sing?

Topic
Homes

Resources
- Book: *Noisy* by Shirley Hughes (Heinemann Library) or *Noisy Book* by Rod Campbell (Blackie Babies)
- Poems: *Noisy Poems* illustrated by Debi Gliori (Walker Books)

- How would it feel to be deaf? Put your finger in your ears. Can you hear what I say?
- Talk about lip reading and sign language.
- Make up sign language to communicate.
- Teach a few simple words in Maketon sign language.
- Talk about those who cannot speak or have difficulty making themselves understood.

Links to home
- Ask any parents (or teachers) who can sign to give a demonstration.

Related activity
- Stranger danger (see page 28)

Angry

Topic
Myself

Resources
- Book: *Angry Arthur* by Hiawyn Oram (Random House)
- Books: *Why Lose Your Temper?* and *Angry* by Janine Amos (Cherry Tree Books)
- Web site: www.mind.org.uk
- Soft play equipment, trampoline
- Cardboard boxes
- Scraps of wood, hammer, nails

⚠ Supervise the use of hammer and nails and outdoor activity.

Learning objective
- To express own needs and feelings in appropriate ways, while remaining aware of how behaviour and actions can affect other people

What to do
Circle time
- Read *Angry Arthur* or any of the alternative books.
- Initiate discussion with the children using the following questions:
  - ◆ What makes you angry?
  - ◆ What do you do when you are angry?
  - ◆ Why do you do that?
- Listen to the children's comments. Empathize with their problems. Share their feelings with the rest of the circle.

- Continue the discussion, asking the children to think about how to cool down instead of staying angry. Encourage suggestions.
- Suggest STOP – THINK – DECIDE.
- Would it help if children who are angry had a 'time out' space in the setting?
- Talk about what action they must not take, and the consequences if they do.
- Remind the children that misusing toys, hitting and throwing things can hurt people and shouting or using rude words is only likely to make the situation worse.
- When there is conflict between children, ask them what happened and give them time to explain. Calm the situation by not appearing to be angry.

Outdoor activity
- Encourage boisterous play by using the soft play area or trampoline.

Indoor play
- Encourage the children to build a tower of cardboard boxes, then squash them.
- Encourage the children to do some woodwork and knock in nails.

Links to home
- Check whether there is disruption at home. Any worries about a child should be shared with parents. Any referrals need to involve the parents.

Related activities
- Don't swing on the curtains (see page 50)
- It's mine (see page 44)
- Sorry (see page 53)
- Laughing (see page 37)
- Whoops-a-daisy! (see page 54)

Time to stop playing

Learning objectives
- To work as part of a group, understanding boundaries and behavioural expectations
- To talk about consequences of not following safety rules and to understand how and why to tidy up.

What to do
Circle time
- Read *I'm Not Your Friend*.
- Initiate discussion with the children, using the following questions:
  - ◆ Why did the mummy fox want to stop playing? (It was getting dark.)
  - ◆ Did the little fox do as he was told?
  - ◆ Does your mummy ask you to stop playing, and you don't want to?
- Have you ever said to your mummy, or to your friend, that you were not being their friend?
- Was it because you wanted your own way?
- Was the mummy fox right to stop their play together? (The little fox became frightened in the dark.)
- Should you play out on your own in the dark?

Activities
- Ten minutes before the end of the session, stop the children playing, and ask them to clear the toys away.
- Ask the children to sit on the carpet in a circle. Discuss whether they were like the little fox and did not want to stop playing.
- Discuss why they needed to stop playing when the practitioner wanted them to.

Links to home
- Ask parents to encourage their child to tidy up after their playtime at home.

Related activities
- New friends (see page 39)
- Stranger danger (see page 28)

Topic
Toys

Resources
- Book: *I'm Not Your Friend* by Sam McBratney (Picture Lions)

Don't swing on the curtains

Topic
Homes

Resources
■ Book: *Big Bears Can* by David Bedford (Little Tiger Press)
■ Soft play area or trampoline

Learning objectives
● To be aware of boundaries set and behavioural expectations at home, at the setting and in other places
● To talk about behaviour and its consequences and know that some behaviour is unacceptable
● To understand that care needs to be taken with property
● To move confidently in a range of ways, safely negotiating space

What to do
Circle time
● Read *Big Bears Can* and then initiate a discussion with the children, asking the following questions:
  ◆ Who was being naughty? (Little Bear was encouraging Big Bear to be naughty.)
  ◆ Should you play in your house so boisterously that you break things?
  ◆ Why did Big Bear break things in the house? (Too big.)
  ◆ Did he do it on purpose? (No, but he should have realized that what he was doing would break things.)
  ◆ What did Big Bear have to do? (Clear up the mess.)

● Talk about the children's experiences of not taking enough care of things in the home, and the financial consequences.
● Where is the best place to be boisterous and bounce about?
● Talk about their experiences on a bouncy castle.

All activities
● Talk to the children and ask whether they are playing carefully with the toys.
● Ask how they can keep the toys safe, so that they don't get lost, damaged or broken.
● Encourage them to think about what would happen if someone encouraged them to do something that they knew would cause damage. They should decide whether they would join in or be able to refuse.

Soft play area
● Enjoy a safe and boisterous bounce.

Related activities
● Sorry (see page 53)
● Jamela's dress (see page 55)

Graffiti

Learning objectives
- To talk about own and others' behaviour and its consequences and know that some behaviour is unacceptable
- To understand that actions affect other people
- To show care and concern for the environment

What to do

Circle time
- Read the book, *The Time It Took Tom*.
- Talk about when the children's parent(s) decorated a room.
- Ask whether the children think that Tom has done the wrong thing by painting everything in the room.
- Did his mummy think that it was funny?
- Ask the children whether they have been told off for writing or drawing on a wall.
- Discuss why graffiti is unsightly.

Art activity
- Paint a picture or mix a piece of baked playdough using a tint, shade and/or tone of one colour.
- A tone is made when white + black + colour are mixed, eg tan, beige.
- A shade is made when colour + black are mixed, eg brown, maroon, olive.
- A tint is made when colour + white are mixed, eg lavender, pink, peach.

Role-playing
- Encourage a role-play with a mother telling her child off for drawing on the walls in his/her bedroom. The practitioner can take the part of the child to extend the activity.
- Puppets can be used for the role-play.

Links to home
- Ask parents to reinforce that it is wrong to draw graffiti.

Topic
Colours

Resources
- Book: *The Time It Took Tom* by Nick Sharratt and Stephen Tucker (Scholastic Hippo Books)
- Playdough
- Paper, brushes, paint – primary colours, black and white
- Hand puppets of a family

Only one sweet wrapper

Topic
Food and shopping

Resources
- Pack of sweets with cellophane wrappers
- Book: *Why Can't I Just Eat Sweets?* by Sally Hewitt and Ruth Thomson (Belitha Press) also as a Big Book
- Board game: *What's Rubbish?*
- Large tins (eg catering tin of coffee)
- Materials to decorate the tin
- Clear sticky-backed plastic

Learning objectives
- To talk about behaviour and its consequences and know that some behaviour is unacceptable.
- To understand that actions affect other people
- To show care and concern for the environment, considering the consequences of dropping litter

What to do

Circle time
- Unwrap a sweet, and drop the wrapper on to the floor. Repeat, until a pile of sweet wrappers is formed.
- Ask the children whether they drop their sweet or lolly wrappers on to the ground.
- Ask what would happen if everyone dropped their litter on the ground, and no-one swept it up.
- Ask where they should put their wrappers.

Music activity
- To the tune of 'Ten Green Bottles', sing
  One sweet wrapper lying on the floor,
  One sweet wrapper lying on the floor,
  And if another sweet wrapper should accidentally fall
  There'll be two sweet wrappers lying on the floor...
- Encourage the children to make up their own song, eg
  One drinks can ...

Book corner
- Support reading non-fiction book, *Why Can't I Just Eat Sweets?*

Table activity
- Play board game, *What's Rubbish?*

Design and making activity
- Make a litter bin by decorating a large tin.
- Make 'stained glass' effect windows by sticking the cellophane wrappers on sticky-backed plastic.

Links to home
- Ask parents to talk about recycling and take their children to recycling banks with their glass, paper and plastic bottles.

Sorry

Learning objectives

- To understand that actions affect other people and to be aware of boundaries and behavioural expectations
- To talk about behaviour and its consequences and to know that some behaviour is unacceptable

What to do

Circle time

- Read 'Harvey the Skateboarding Octopus'. Harvey steals from his mummy because he wants to buy a skateboard.
- Initiate further discussion with the children, using the following questions:
  - ◆ Discuss whether it was right to take the money.
  - ◆ How would you feel if someone had taken one of your toys?
  - ◆ Did Harvey feel better for owning up and saying sorry?
  - ◆ Do you think that Harvey really meant that he was sorry? How do you know? (He won't steal again.)

Role-playing

- Encourage the children to use the hand puppets to talk through a scenario involving breaking a vase at home, and owning up.

Book corner

- Support reading the books in the Resources list.

Topic
Homes

Resources

- Books: Growing Up series: *Owning Up* by Janine Amos (Cherrytree Books)
- Books for practitioner and older children: Choices series: *Telling the Truth* (A&C Black)
- Hand puppets
- Book: 'Harvey the Skateboarding Octopus' from *The Reluctant Mole and More Beastly Tales* by Philip Welsh (Scripture Union)

Related activities

- How thoughtful (see page 69)
- Be kind (see page 70)
- Thank you (see page 71)
- Me first (see page 45)
- Hello (see page 65)

© Mavis Brown and Rebecca Taylor and Brilliant Publications

Personal, Social and Emotional Development with Understanding the World and Mathematics

Whoops-a-daisy!

4

Topic
People who help us

Resources
- Nursery rhymes: 'Humpty Dumpty', 'Hickory Dickory Dock', 'Ring o' Roses', 'Incy Wincy Spider', 'Up the Tall White Candlestick' from *This Little Puffin* compiled by Elizabeth Matterson (Puffin Books)
- Book: *Ragged Bear's Book of Nursery Rhymes* illustrated by Diz Wallis (Ragged Bears)
- First Aid kit with bandages
- Large teddies and dolls
- CD of nursery rhymes
- CD player
- Large jigsaw puzzles

 Supervise sensible use of bandages.

Learning objectives
- To respond and show sensitivity to the needs and feelings of others
- To talk about how people show feelings in stories, using the misfortunes of nursery rhyme characters
- To listen to and join in with stories and rhymes, accurately anticipating key events and responding to what is heard with relevant comments and questions

What to do
Playtime
- Comfort a child when she falls over, find out how the injury happened and encourage other children to support the child.
- Talk about nursery rhyme characters who fall over, fall off or tumble down.

- Suggest that the children could play at bandaging the teddies in the role-play area.

Literacy activity
- Listen to the nursery rhymes on the CD.

Role-playing
- Using dolls or puppets, sing or say together with the actions 'Jack and Jill Went Up the Hill'.
- Ask the children whether they think it was a good idea to wrap Jack's head in brown paper and vinegar and explain the theory behind it and the modern day equivalents (plasters, creams and ointments).
- Sing or say together 'Humpty Dumpty'.
- Ask the children how they would try to put Humpty together again.
- Play at wrapping up the toys' broken arms and legs.

Table activity
- Encourage children to play with the jigsaws (put Humpty together again).

Links to home
- Talk to parents about any accidents and injuries and always ask them to sign the appropriate accident/incident forms for the setting.

Related activities
- The tree house (see page 32)
- Ring games (see page 72)
- Laughing (see page 37)
- Angry (see page 48)

Jamela's dress

Learning objectives
- To understand that actions affect other people
- To talk about behaviour and its consequences and know that some behaviour is unacceptable

What to do
Circle time
- Read *Jamela's Dress*.
- Initiate discussion with the children, using the following questions:
  - What did Jamela's mummy ask her to do? (Look after the fabric drying on the washing line.)
  - What did the townspeople think of Jamela walking down the street with her mummy's dress material?
  - What did the photographer do?
  - How did Jamela's mummy feel when she saw the ruined dress material?
  - Did Jamela mean to spoil the fabric?
  - What happened next?
  - How did Jamela feel?
  - Why was everyone happy at the end of the story?
- Have you done something that you were sorry about later?

Water tray
- Wash pieces of fabric and hang them out to dry.

Extensions/variations
ICT activity
- Show the children how to take photographs.
- Encourage them to take photographs of each other.

Related activity
- Sorry (see page 53)

Topic
Water

Resources
- Book: *Jamela's Dress* by Niki Daly (Frances Lincoln)
- Water tray: warm water, soap, pieces of fabric
- Pegs, string
- Single-use camera or digital camera

 Follow local authority and setting guidelines on the use of photographs of the children.

The four Cs

Topic
Shapes

Resources
■ Book: Lifetimes series: *Mother Theresa* (Chrysalis)

Learning objectives
● To understand that there needs to be agreed values and codes of behaviour for groups of people, including adults and children, to work together in harmony
● To talk about own and others' behaviour and its consequences and know which types of behaviours are desirable and which are unacceptable

What to do
Circle time
● Tell the children that you want them to think of some rules that everyone should follow, so that everyone can play and work happily.

● Remind them that the rules apply to both adults and children. Ask what they think should be done if the rules are broken.
● There are four words that begin with the letter C that can help us to behave well.
● They will be referred to as 'the four Cs'.
● *Cooperation* means working together. (Good examples would be parachute games, performances and group art displays.)
● *Consideration* means thinking of others, taking responsibility for actions and behaviour and thinking of how it affects those around you.
● *Compassion* means feeling sympathy for the sufferings of others. (Someone who cries when a story or film is sad shows that they have compassion.)
● *Caring* means looking after other people, possessions, property or the environment.
● All of these qualities are very important and contribute to a happy community.

Links to home
● Explain to parents that the children have learned about 'the four Cs' and what they are.

Related activities
● How thoughtful (see page 69)
● Be kind (see page 70)
● Thank you (see page 71)
● Round and round (see page 73)
● Me first (see page 45)
● Sorry (see page 53)

Can't play outside

Learning objectives
- To adjust behaviour to different situations, be flexible and accept changes in routine
- To begin to be able to negotiate to solve problems

What to do
Circle time
- Outside, the weather is raining/snowing/ very windy. Change the weather chart.
- Initiate a discussion with the children as a group, asking the following questions:
  - What's the weather like today?
  - Should we go outside to play?
  - What will happen to us if we do?
  - What can we do with our time if we don't go outside to play?
- Encourage the children to choose an activity for the whole class.
- Encourage deeper thinking by asking the following questions:
  - Why is it important to be able to change what you are planning to do?
  - Is it easier to change things on your own, or when working in a group?
- Ask the children to consider their choice carefully by thinking:
  - If the group wish to watch a DVD, negotiate which DVD they want to watch.
  - Will it teach us about the topic we are learning?
  - Will most of us enjoy it?

Topic
Seasons/Weather

Resources
- Chart showing 'Today's Weather'
- DVD: as requested
- DVD player and television

  - Were you away from the setting when the group saw it the first time?
  - Do you want to watch it again?

Related activity
- All the fun of the fair (see page 63)

Personal, Social and Emotional Development with Understanding the World and Mathematics

Personal, Social and Emotional Development

Making Relationships

- Children who make friendships easily are usually those who have had positive early experiences and strong relationships with their adult caregivers. In order to be socially confident, they need to feel emotionally, as well as physically, safe.

- The ability and desire to form positive relationships is dependent upon the knowledge that there is enough love and attention for everybody. Children should never feel that they have to vie with others for their share or be better than everybody else in order to win approval.

- Being encouraged to empathize with others means that children are less likely to hurt their friends or spoil their games, because they understand how it feels to be treated badly. They should feel able to make mistakes, to keep trying despite obstacles and setbacks, to help and encourage other children and to be realistic in their expectations. They should learn how to ask for what they want, how to take account of others' needs, feelings and opinions and when to say 'thank you' or 'sorry'.

- If children know that their own thoughts and opinions are valid and will be respected, they will feel less need to dominate and will not always expect to have their own way.

- When children have gained enough experience of positive relationships, they will feel able to seek out and make new friendships and to socialize and cooperate constructively with others.

- Practitioners must provide an environment that supports these developing abilities and the security of key people to return to for comfort and encouragement when things go wrong or life feels threatening. This will enable children to build essential skills at their own pace and to become personally, socially and emotionally confident.

Tea time

Learning objectives
- To explain own knowledge and understanding and ask appropriate questions of others
- To play cooperatively, taking turns with others and taking account of others' ideas about how to organize an activity
- To handle tools and malleable materials safely and with increasing control
- To engage in imaginative role-play based on first-hand experiences

Preparation
- Practitioner to make playdough, colouring batches with different food colouring.

What to do
Circle time
- Read *The Tiger Who Came To Tea*.
- Talk about when the children take their meals. Use their daily routine chart to point out snack time, and, if appropriate, lunch time.
- Ask what the children eat at tea time, dinner time or supper time at home.
- Support the learning of new vocabulary and developing language skills by introducing words such as: *sandwiches, cake, tea, milk, ham, cheese, pizza, chips, potatoes, rice, pasta, meat, fish, carrots, beans* and *yoghurt.*

Craft activity
- Encourage the children to make pretend food with the playdough.
- Talk about how they are making the food.
- Support the learning of new vocabulary and developing language skills by introducing words such as: *roll, cut, press, smooth, rough, round, thin* and *flat.*
- The baked dough can be painted with poster paints mixed with a little PVA glue.
- Use for role-play.

Extensions/variations
Role-playing
- In the role-play area, encourage the children to lay tables, serve and clear away.
- Observe whether a child makes room for another child at the table.

Topic
Food and shopping

Resources
- Book: *The Tiger Who Came To Tea* by Judith Kerr (Picture Lions)
- Chart showing a daily routine with meal times
- Playdough
- Paints, paintbrushes
- PVA glue
- Rolling pins, plastic knives
- Tools to shape and mark playdough (not sharp)
- Role-play area set out as dining room or café; plastic knives, forks, spoons, small plates, tea cups and saucers, large plates for food

Links to home
- Ask parents to encourage their child to help lay the table, and introduce 'left' and 'right'.

Related activities
- Honey cake (see page 24)
- At the café (see page 60)

Personal, Social and Emotional Development with Understanding the World and Mathematics

At the café

Topic
Food and shopping

Resources
- HCD-ROM: At the Café (SEMERC)
- Computer
- 10 minute sand timer
- Role-play area set out as a café with table, tablecloth, plates, cups and saucers, teapot, sugar bowl, milk jug, cutlery
- Playdough food
- Aprons for children
- Notepad and pencil
- Cash register and toy money

Learning objectives
- To play cooperatively, sharing resources and taking turns to use equipment
- To be sensitive to the needs and feelings of others and to form positive relationships and friendships with other children
- To complete a simple game or program on the computer, within a small group
- To engage in imaginative role-play based on own first-hand experiences
- To understand that all games, toys and future career options are open to both genders equally

Preparation
- Prepare the role-play area as a café.
- Load the CD-ROM.
- Help the children make pretend food from playdough.

What to do
Circle time
- Invite children to take turns to play in the role-play café and with the computer game, in small groups, choosing when to ask for a turn and inviting friends to join them.
- Tell them that you will be noticing whether everyone in the group is being polite and saying 'Please' and 'Thank you'.

Role-playing
- Support children and join in with their role-play as they lay tables, order and serve food and drinks and tidy up, taking turns to be chefs, waiters and waitresses and customers.
- Ask them whether there are any cafés near to their homes or in the shops that they visit with their parents.
- Talk about any times that they have been inside a café, what they ordered to eat or drink and if it was expensive.

Computer activity
- Ensure that the children are sharing and taking turns. They can use the sand timer.
- Help them to form a strategy for taking turns.

Related activities
- Tea time (see page 59)
- Honey cake (see page 24)
- Me first (see page 45)

Let me introduce myself

Learning objective

- To demonstrate friendly behaviour, initiate and respond to conversations and form positive relationships with peers

Preparation

- Address A4 envelopes so that each child has an envelope labelled 'From (child's name)'. Ask each child to draw a picture of herself on an A4 sheet of paper and put it inside her envelope.
- If there is an odd number of children in the group, a practitioner can draw a picture of him/herself, or ask one child to draw two pictures of herself for two envelopes.

Topic
Myself

Resources
- Addressed A4 envelopes
- A4 sheets of paper
- Selection of board games and small-world toys

What to do

Circle time

- Gather the children to sit in a circle and invite them to exchange their envelopes with the children opposite them.
- The giver must say, 'Hello, my name is X. This is for you.'
- The receiving child must say, 'Thank you X,' then give their picture saying, 'My name is Y. This is for you.'
- Make sure that each child has received a picture.
- Make a record of the givers and receivers.

Mark-making activity

- Each child should write a thank you note for their picture.
- The practitioner can scribe from a dictation, or the child can make marks, and tell the practitioner what it says.

Table or small-world activity

- At the beginning of an activity suitable for pairs of children to play together, the thank you note is given to the child saying, 'Thank you for the very nice picture, X/Y.'
- Encourage the pairs of children to play with or alongside one another.

Related activities

- My friend (see page 62)
- New friends (see page 39)
- Hello (see page 65)

Personal, Social and Emotional Development with Understanding the World and Mathematics

Making relationships

My friend

Topic
People who help us

Resources
- Poem: 'Bernard' from *Rhymes for Annie Rose* by Shirley Hughes (Red Fox)
- Flip chart and marker pen
- A3 sheets of paper
- Paints and paintbrushes
- Pencils

Learning objectives
- To demonstrate friendly behaviour, initiate and respond to conversations and form positive relationships with peers
- To handle tools effectively, including paintbrushes and pencils for drawing

What to do
Circle time
- Read the poem 'Bernard', and show the pictures.
- Ask the children why Annie Rose and Alfie like Bernard.
- Invite children to say what it is that they like about their friend or any child (or adult) in the group.
- Write down what they say on the flip chart.
- Ask children if they like some of the same things (what they have in common). Invite them to describe these similarities. For example: *My best friend is Josh. He likes playing on bikes, and I like playing on bikes too.*

- Ask the children what it is that makes a good friend. Suggest that good friends help each other and having a friend can make you feel good about yourself.
- Suggest that some people may say that they are friends, but then let each other down. For example, they may not help each other when they are asked to.

Art activity
- Invite the children to paint and draw pictures of themselves playing with friends.
- Encourage them to talk about what is happening in their pictures as they paint.

Display
- Mount the pictures with the children's comments written down earlier.

Related activities
- Let me introduce myself (see page 61)
- Be kind (see page 70)
- New friends (see page 39)
- Hello (see page 65)

All the fun of the fair

Learning objectives
- To play cooperatively, taking turns with others
- To show sensitivity to and consideration for other people's needs and feelings
- To select and use activities and resources independently or with some support
- To handle objects and construction materials with increasing control

Preparation
- Find the story of 'The Fair' on the DVD.
- Load the CD-ROM of *Touch Funfair* on the computer.

What to do

Circle time
- Show the DVD of Kipper at the fair.
- Talk about all the different rides in a fair.
- Talk about going round in circles.
- Support the learning of new vocabulary and developing language skills by introducing words such as: *tipping, spinning, rotating, rising* and *falling*.
- Talk about things the children do not like about fairs, eg goldfish given away as prizes, rubbish left on the ground, noise.
- Encourage the children to choose an activity, and remind them to take turns.
- Note which children persist with one activity at a time and which children visit several activities but do not carry out any to their conclusions.

Outside activity
- Throw beanbags at skittles.
- Encourage number work by asking the children how many they knocked over or how many were left standing.

Design and making activity
- Encourage cooperative play while making the models.
- Make a merry go round or swing, big wheel, plane ride or train ride using the construction toys.
- Make the *Marble Run*, and allow other children to try it out.

Small world activity
- Play 'bumper cars' with the toy cars inside a hoop laid on the floor.

Topic
Shapes

Resources
- DVD player and television or computer
- DVD: *Kipper – The Big Freeze and Other Stories* select 'The Fair'
- Computer
- CD-ROM: *Touch Funfair* (SEMERC)
- Skittles, bean bags
- Construction toys
- Toy: *Marble Run*
- Small toy cars and hoop
- Game: *Snail's Pace Race*
- For reference: www.kipperthedog.com

- When children's cars approach each other, suggest that they say 'Excuse me' and, if cars bump, encourage the children to say 'Sorry'.

Table activity
- Set up the game *Snail's Pace Race*.
- Ensure that children play fairly at the board game, even though they may play to their own rules.

Computer activity
- Ensure there is turn-taking when playing with the computer game.

Links to home
- Use some activities as part of a fund-raising fête at the setting.

Related activities
- Sports day (see page 26)
- First visits (see page 18)

Welcome to my nursery

Topic
People who help us

Resources
■ Multi-lingual 'Welcome' posters with name of setting
■ A3 paper and crayons

Learning objectives
● To explain own knowledge and understanding and attend to and take account of what others say
● To work and play cooperatively, taking turns and sharing with others
● To form positive relationships and friendships with adults and other children
● To describe self in positive terms and understand the value of being a member of the group and the setting
● To represent own ideas, thoughts and feelings through art

Preparation
● If the setting has a number of children, provide stability by keeping children in the same working groups.

What to do
Circle time
● If the group has children with English as an alternative language, ask them to say 'Hello' in their home language.
● Share the posters with the children and initiate discussion, considering such questions as:

◆ Has this group got a name?
◆ What are the names of the adults in the group and at the setting?
◆ Who do you play with when you come here?
● Encourage children to talk about the setting, the practitioners and their friends, offering support as necessary. For example, they might say, 'I go to play with Jane.'

Art activity
● Provide paper and crayons for all children.
● Explain that you would like to make pictures of the setting and invite the children to choose whether to draw them inside or outside.
● After one minute of drawing, they pass their paper to the child on their right, who adds to the picture.
● Stop when the pictures return to their original artist.
● Talk about how they all helped to draw the pictures.
● Display their work.

Related activity
● First visits (see page 18)

Hello

Learning objectives
- To demonstrate friendly behaviour and initiate conversations
- To have a positive self-image and be confident to speak in a familiar group
- To form positive relationships with adults and peers

What to do
Circle time
- Sit in a circle.
- A child says to the child on their left, 'Hello, my name is Jane.'
- Second child says to child on their left, 'Hello, my name is Nathan, and this (indicating child on their right) is Jane.'
- Encourage children to continue with the activity around the circle until each child has spoken. Practitioners can sit within the circle and join in. This is an ideal way for everybody to learn each other's names.
- Saying 'Hello' helps to make friends.
- Share the poster with the children and ask them if they know of any different ways of greeting people.
- Tell the children about different forms of greeting in other cultures and historical times, eg bowing and curtseying to the Queen; raising a hat; shaking hands; the French kiss on both cheeks; the Inuit (Eskimos) rub noses.
- Learn to say 'Hello' and 'Goodbye' in different languages.

Role-playing
- Role-play different situations meeting:
  - the Queen
  - two friends who haven't seen each other for years
  - someone you don't know very well
  - your grandma
  - two people who don't like each other.

Topic
Myself

Resources
- Multi-lingual Welcome poster

Related activities
- Let me introduce myself (see page 61)
- My friend (see page 62)

Making a secret garden

6

Topic
Gardening

Resources
- Bulbs, plants, shrubs, climbing plants and trees
- Trowels
- Spades and forks for adults
- Small plant pots, large plant pots, compost and water-retaining gel
- Calendar showing months to record progress of growth
- Camera

⚠ Wash hands after working with soil or compost. Check that the children have had tetanus injections before working in the garden.

Learning objectives
- To explain own knowledge and understanding and ask appropriate questions of others
- To work and play cooperatively, taking account of one another's ideas about how to organize an activity.
- To demonstrate a sense of pride in own achievements and welcome and value praise from adults
- To safely use and explore simple tools and techniques

Notes for practitioner
- Choose the following types of plants:
  - spring and summer bulbs that flower across a period of time, but do not require staking, eg snowdrops, crocus, daffodils, narcissus, early flowering tulips, late flowering tulips, Dutch irises, hyacinths

- fragrant plants: lavender, lily of the valley, dianthus, hyacinth
- interesting shapes: *Artemisia stelleriana* (jagged silver leaves), witch hazel (fragrant early flowering shrub), passion flower (unusual flowering climber, but needs sun and no frost), tamarisk (fragrant shrub with thin leaves), *Picea pungens Nana* (dwarf blue conifer tree), umbrella grass *Cyperus alternifolius* (decorative grass)
- attractive to insects: honeysuckle (fragrant climber, attracts bees), buddleia (attracts butterflies)
- shrubs to train into tunnels: dogwood (grows autumn/winter long thin red twigs), forsythia (yellow spring flowers)

- Prickles and thorns are interesting, eg holly, but select the position of the bush carefully.
- Berries can be too tempting for small children, so do not include berry-producing plants. Laburnum pods are very poisonous. Euphorbia sap is caustic.

What to do
Outdoor activity in garden
- Plant the small plants and bulbs in the agreed areas of the garden.
- Support the learning of new vocabulary and developing language skills by introducing words such as: *plant, compost, bulb, crown, roots, stem, flower* and *leaf.* Show which way up to plant the bulbs, and how deep into the ground.
- Record the growth on the calendar.
- Encourage children to take photographs of their plants at intervals during the growing period and then to make a display showing 'Before' and 'After'.

Links to home
- Ask for help with the planting of large plants and erection of any fences, gates and archways.

Related activity
- Stepping stones (see page 36)

Harvest basket

Learning objectives

- To demonstrate friendly behaviour, initiating conversations confidently and forming good relationships with adults
- To show interest in the lives of unfamiliar people
- To manipulate materials to achieve a planned effect
- To use simple techniques competently and appropriately
- To experiment with colour and design

Preparation

- Photocopy and enlarge harvest basket templates (see page 190).
- Harvest baskets can be distributed to the elderly in the community. Make contact with a local retirement home or the warden of a local sheltered housing complex.
- The children could also sing to the residents.

What to do

Circle time

- Explain to the children that they are going to take gifts of food including sweets and biscuits to the elderly people of the area.
- Suggest that they make and decorate a basket to carry the food.
- Discuss what the children might say to the people.

Craft activity

- Cut out the template along the thick lines.
- Fold along the thin lines.
- Decorate the outside of the basket (but not the base or flaps).
- Stick the handle long side across the inside of the base and the side of the basket, as indicated.
- Fold the basket, with flaps being stuck inside or stapled over the handle.

Topic
Celebrations/Gardening

Resources

- Template of harvest basket on thick card (page 190)
- PVA glue and glue spreaders
- Stapler
- Materials to decorate the basket – the child's choice
- Scissors
- Sweets or biscuits (some for diabetics) for small basket
- Songs and CD from *Start with a Song* by Mavis de Mierre (Brilliant Publications)
- Songs from *This Little Puffin* compiled by Elizabeth Matterson (Puffin Books)
- Children's hymns

> ⚠ Supervise the use of scissors and stapler.

Music activity

- Teach the children some songs that would be appropriate to sing at harvest time.

Links to home

- Ask parents if they could accompany their children while they distribute the harvest baskets.

Take a message

Topic
People who help us

Resources
- Book: *Don't Forget the Bacon* by Pat Hutchins (Red Fox)
- Examples of messages on different media, eg handwritten letter, word processed letter, faxed message, Braille (there are Braille dots on prescribed medications and children could investigate them on the empty boxes)
- Telephones
- Computer with modem, connected to telephone line

Learning objectives
- To initiate conversations, listen attentively, attend to and take account of what others say and respond appropriately
- To explain own knowledge and understanding and ask appropriate questions of others
- To work and play cooperatively
- To follow instructions involving several ideas or actions
- To express messages effectively, showing awareness of listeners' needs

What to do
Circle time
- Read book *Don't Forget the Bacon*.
- Tell children that you sometimes forget what people say to you and ask whether they do too.
- Play 'Chinese Whispers'. Pass round a simple whispered message.

- Discuss whether anyone made up a message because they could not hear what was said.
- Ask children what they should do if they do not hear or do not understand messages that people give to them. (Ask them to repeat it.)
- Encourage small groups of children to take simple spoken messages to each other.
- Talk about the different ways in which a message can be sent. Support the learning of new vocabulary and developing language skills by introducing words such as: *letter, telephone, fax, e-mail, mobile phone, text.*
- Talk about how messages were sent in the past: *spoken by a messenger, lighting a beacon, ringing church bells, telegraph, Morse code, flags on ships, semaphore.*
- Talk about sending messages to blind or deaf people: *Minicom, Braille.*

Extensions/variations
- Play at speaking on the telephone.
- Send an e-mail to a friend or family member, or to a children's television programme.
- Send a message to Santa.
- Give simple instructions to the youngest children, such as, 'Go and fetch ... , for me, please.'

How thoughtful

Learning objectives
- To show sensitivity to others' needs and feelings and to form positive relationships with adults and other children by being helpful and thoughtful
- To adapt behaviour to different events and social situations
- To understand that own actions affect other people
- To talk about how people show feelings and the consequences of different behaviours

What to do

Circle time
- Read *Tidy Up Titch* and talk about why it was a good thing for Titch to tidy up. Explain that it makes children and adults feel happier if everybody pleases everybody else by being helpful and thoughtful. Initiate more discussion with the children, asking the following questions:
  - Can you think of other ways in which you can help your mummy or daddy at home?
  - Can you think of any ways you can help the practitioner, other children, or other adults?
  - Include ideas, such as: opening doors, or standing back to let grown ups through first
  - waiting while mummy finishes talking before you talk (or put up your hand when at the setting)
  - helping each other to solve problems;
  - volunteering to be a monitor and doing the job well
  - finding, using and returning materials for yourselves; tidying up after an activity
  - handing out snacks etc.

Class
- Listen carefully to their replies, adding supportive comments to their suggestions, and be seen to write down their ideas on the chart for display.

Topic
Homes

Resources
- Book: *Tidy Up Titch* by Pat Hutchins (Red Fox)
- Star stickers
- White board or flip chart and marker pen

Display
- Make a display on 'I have been helpful today' with their suggestions and a drawing to illustrate the idea and a space for children's names.
- When a child performs one of the helpful deeds suggested, write her name on a star and put it beside the appropriate drawing on the display. Allow her to take the star home at the end of the session.
- Make a 'monitors' list. Draw a symbol indicating a task, eg sweeping brush for the sweeping up monitor, books for tidying up the book shelf/trolley. Add children's names and swap them regularly.

Links to home
- Ask parents to encourage their children to be helpful and to tidy up their toys at home.

Related activities
- Well done (see page 38)
- Be kind (see page 70)
- Thank you (see page 71)
- Me first (see page 45)

Be kind

Topic
Myself

Resources
- Book: *Three Cheers for Ostrich* by Francesca Simon (Gullane Children's Books)
- White board or flip chart and marker pen
- Book: Any *Mr Men* or *Little Miss* stories by Roger Hargreaves (World International)
- CD with songs: *Mr Men* and *Little Miss* (Delta) www.deltamusic.co.uk
- CD player
- Smiley stickers:

Learning objective
- To show sensitivity to others' needs and feelings and to form positive relationships with adults and peers by being kind

What to do
Circle time
- Read the book *Three Cheers for Ostrich*.
- Talk about what each animal can do. (The cheetah is fast, the elephant is strong.)
- Invite children to say positive things about each other. Encourage each child to contribute an idea about someone else.
- Talk about being kind to each other. Explain that it makes everyone feel much happier.
- Ask the children what kinds of things the animals said that Ostrich had done for them. (Ostrich looked where he was going etc)

Extension/variation
- Read *Mr Men* and *Little Miss* stories with the children and listen to their songs, to illustrate a variety of characteristics, attributes and behaviours and encourage use and understanding of descriptive language.

Display
- Make a display entitled 'I Have Been Kind Today' with spaces for children's names and what kindness they did to be placed alongside on paper.
- A smiley face sticker could be given to a child to take home at the end of a session when her name has been put into the display.

Related activities
- Well done (see page 38)
- How thoughtful (see page 69)
- Thank you (see page 71)
- The four Cs (see page 56)
- Me first (see page 45)
- Sorry (see page 53)

Personal, Social and Emotional Development with Understanding the World and Mathematics

Thank you

Learning objectives
- To demonstrate friendly behaviour, initiating conversations and forming good relationships with familiar adults
- To show interest in and appreciation for the different occupations and ways of life of 'people who help us'
- To give meaning to marks that they make as they draw and write
- To attempt to use identifiable letters or other marks to communicate meaning and to write own name

Preparation
- Ensure that resources are accessible to the children.
- Invite non-teaching staff who work at the setting, such as cleaners, caretakers, cooks and office staff, to talk to the children during a session and demonstrate some of the work they do.

What to do
Circle time
- Talk with the children about who works at the setting and make notes on the chart.
- Ask the children if they understand and can explain what the different people do.
- Walk around the setting with small groups of children so that they can watch what other people are doing, such as the cook making lunch in the kitchen or the manager speaking on the telephone in the office.
- Encourage the children to ask questions.

Art activity
- Encourage children to make thank you cards for people who work or help at the setting and to write their names inside. Share the poster with them. If the first languages of any of the adults are represented on the poster, support the children in trying to copy the correct letters or scripts into cards for them.
- Some children may prefer to make a card for a family member or friend.

Topic
People who help us

Resources
- Poster: *Thank You* multi-lingual
- Flip chart with marker pen
- Book: Good Manners series: *Thank You* by Janine Amos (Cherrytree Books)
- Computer, colour printer and card for typing the message
- Materials to decorate card
- Card
- Scissors
- Pens, pencils and crayons

Related activities
- Stop look listen (see page 41)
- How thoughtful (see page 69)
- Be kind (see page 70)
- I want to be … (see page 27)
- The four Cs (see page 56)
- Me first (see page 45)
- Hello (see page 65)

Personal, Social and Emotional Development with Understanding the World and Mathematics

Ring games

Topic
Shapes

Resources
- Songs: 'Ring-a-Ring-o-Roses', 'Here We Go Round the Mulberry Bush', 'The Farmer's in His Dell', 'There Was a Princess Long Ago', '(Sally) Go Round the Sun', 'Row Your Boat' from *This Little Puffin* compiled by Elizabeth Matterson (Puffin Books)
- Books: Books with Holes series (Child's Play)

Learning objectives
- To play cooperatively within a group, taking turns and sharing fairly, to form positive relationships with adults and other children
- To confidently try new activities and display a positive sense of self and own abilities
- To understand that own actions affect other people and adjust behaviour to fit different situations
- To develop control and coordination in large and small movements

What to do
Music activity
- Learn any of the songs with the actions.
- Decide with the children whether it is a clapping, skipping, dancing, choosing or ball game.
- Ask them why we have to work together and consider each other to make the game work? (If we pull in the wrong direction or, if we are too rough, the game won't be fun.)
- Encourage cooperative play and turn-taking.

- Encourage the children to play the ring games, with and without adult support, during free play times.
- These kinds of singing games help the children to interact as a group and learn each other's names.

Book corner
- Support reading the illustrated song books.

Links to home
- Encourage parents to play the games with their children.

Related activities
- The tree house (see page 32)
- Whoops-a-daisy (see page 54)

Round and round

Learning objectives
- To work and play cooperatively, sharing and taking turns with others
- To work as part of a group, understanding and following the rules of games and adjusting behaviour to fit the situation
- To develop control and coordination in whole body movements

What to do

Circle time
- Remind the children that one of the rules of the group and the setting is to cooperate, which means to work together.
- Explain that more can be achieved when people work together.
- Explain that they are going to have fun with a parachute.

Outside activity
- Warm up by telling the children to make themselves as tall as they can, then curl up to be as small as they can.
- Ask the children to run round and jump as if they are catching bubbles.

Parachute activity
- *Popcorn*: Place a number of small plastic balls on the parachute.
- Shake the parachute to make them rise like popcorn.
- *Rollaball*: Everyone holds the parachute taut. Place a large ball near the edge, and give it a push to get it rolling.
- To get the ball to come to you, you lower the edge of the parachute.
- To push the ball away, you lift your edge.
- Try to get the ball to run round the edge of the parachute. This requires a lot of concentration and cooperation as all the children will need to synchronize their movements.
- Give praise and encouragement.

Topic
Shapes

Resources
- Parachute (365cm diameter)
- 10 small balls
- Large ball

⚠ Take care that over-enthusiasm doesn't strain muscles. Adult support is needed for this activity.

Why won't you play with me?

Topic
Myself

Resources
■ No special requirements

Learning objectives
● To have a developing awareness of their own needs, views and feelings and express them in appropriate ways
● To be sensitive to the needs, views and feelings of others
● To form positive relationships with adults and other children
● To demonstrate friendly behaviour and initiate conversations
● To take steps to resolve conflicts with other children, such as finding a compromise

What to do
Role-play activity
● If a child is upset because she is excluded from a game, activity or group of friends, encourage her to think about why the other children don't want to play with her and what she might say or do to change their minds.
● Ask: Have you upset them? Do you share and play fairly? (They won't want to play if you grab all of the toys.)
● Suggest ways of encouraging them to want to include her in their game. (Say: I'm sorry, please let me play. You can share my ball.)
● Ensure that there is no issue of bullying or discrimination.
● Suggest finding other children who are feeling kinder and asking if they would like to share a game.
● Decide whether there is an acceptable reason for exclusion from a particular game, such as not knowing the rules.
● Offer a compromise, such as asking if the children could let her know when they start a new game, or teach her the rules.

Related activities
● The four Cs (see page 56)
● It's mine! (see page 44)

Personal, Social and Emotional Development with Understanding the World and Mathematics

Understanding the World

People and Communities

- In order to understand the world that they live in, young children must first understand their own place within it and the familiar people who surround them.

- Practitioners should encourage children to talk about both significant events and ordinary everyday life in their homes and with their families, and to share their experiences with the adults in the setting and with each other. They should begin to recognize similarities and differences between people and between families, communities, cultures and traditions and to understand that they are all worthy of equal respect.

- Settings can introduce children to the many different beliefs and lifestyles that they may come into contact with, by celebrating a variety of festivals and special days throughout the year and finding out about the favourite and traditional foods, clothes, animals, plants and climates of countries around the world. Inviting family members into the setting to talk to the children about their cultures, or to share cookery, stories, songs, dances or outfits with the group, is an excellent way of reaching out and making all families feel welcome and involved.

- It is important to celebrate diversity and promote equal opportunities for all children and families using the setting. It may be necessary to make some changes or to provide particular services to enable those of different cultures and religions, those learning English as an additional language and those with disabilities to be fully included.

- Practitioners may need to explain carefully and encourage children to develop an understanding of why some people need extra help and support or a different approach in some areas, or sensitivity when they feel upset by particular things. All children need to be sure that they will be treated as individuals and that their specific characteristics and particular needs will be accommodated.

- Children are interested in people and the ways in which we are all different, but they do not make judgements and will absorb the attitudes of the adults around them. If they do not experience prejudice or discrimination, they will not apply them to others. If they grow up within an accepting and enabling environment, in which people work to accommodate each other's needs and wishes, they will strive to do the same and to always build supportive and constructive communities.

All grown up

Resources
- *When I Was a Baby* by Catherine Anholt (Picture Mammoth)
- Calendar
- Dates of the children's birthdays
- For role-play area, set up as a nursery: baby dolls, larger dolls
- Doll's house
- *On Your Potty* by Virginia Miller (Walker)
- Photographs of the children (from birth to present day)

Learning objectives
- To remember and talk about significant events in own experience
- To talk about own home and community life and to find out about other children's experiences
- To look closely at similarities and differences, patterns and change
- To make observations and develop an understanding of growth and changes over time
- To use everyday language relating to time

What to do
Circle time
- Read *When I Was a Baby*, and discuss each picture. Ask the children to say what they can do now, that they could not do when they were a baby.
- Ask the children to point to their birthday on the calendar. Ask the children who are two/three/four/five to stand up in turn. Talk about being older or younger.
- Point out that the taller children are not necessarily the older ones.
- Support the learning of new vocabulary and developing language skills by introducing words such as: *how old? eldest, youngest, growing, baby, toddler and child*. Make comparisons between then and now, describing events, achievements and milestones in the children's lives.
- Encourage children to try to remember when they first came to the setting, especially those who have attended for up to two years.

Table activity
- Support children as they sequence photographs of their lives from birth to the present day.

Role-play
- Join in when a child is playing with a doll and talk about how old the baby/toddler/small child is, and what it can do. Relate the questions to the story.

Small world activity
- Ask who lives in the house and their ages.
- Ask what the people in the house are doing. Relate the questions to the story to determine the child's understanding.

Extension/variation
- Talk about the story *On Your Potty*, and how grown up the children are when they can go to the toilet on their own.

Links to home
- Ask parents to lend 4–5 photographs of their children at various ages from birth to the present day. Remind them to label the backs of the photographs with the children's names.

Related activities
- What a picture! (see page 130)
- Our baby (see page 88)

Religious stories

Learning objectives
- To talk about own home and community life and find out about other children's experiences
- To enjoy joining in with family customs, routines, faiths and celebrations
- To celebrate diversity, respect the beliefs and values of others and develop a positive attitude to individual needs and differences
- To understand similarities and differences among cultures, communities and traditions, by sharing and discussing practices, resources and experiences
- To develop awareness and understanding of the times of year at which festivals are celebrated and the lengths of time between them
- To listen attentively to stories and respond with relevant comments and questions

Preparation
- Prepare a yearly calendar and write in when the main religious festivals occur.
- The dates of some festivals vary because they are determined by the lunar cycle or sighting of the new moon.

What to do
Circle time
- Look at the calendar to point out the different festivals through the year.
- Read a story appropriate for the time of the year, eg *Samira's Eid*.
- Talk about the significance of the festival.

Major festivals
- March/April: Eid-ul-Adha – Muslim
  Holi – Hindu
  Lent/Easter – Christian
  Passover – Jewish
- September: Jewish New Year
- October/
  November: Divali – Hindu
- December: Hanukkah – Jewish
  Ramadan – Muslim
  Christmas – Christian

Extension/variation
- Role-play one of the stories relating to a festival.

Topic
Celebrations

Resources
- *Samira's Eid* by Nasreen Aktar (Mantra)
- *Celebrations* (Ginn Reading Books)
- *Religion Photopacks* (Wildgoose) – these are for older children but can be adapted
- Stories for Christmas time, eg Lion Children's Books

Links to home
- Explain to parents that there will be an ongoing 'festivals' topic throughout the year. Ensure that all practitioners are aware of the religions that each family follows and which families do not follow any religions.
- Invite parents to come into the setting at appropriate times and talk about their beliefs and how they celebrate particular festivals.
- Ask for help and advice from appropriate parents in advance, to avoid making mistakes that could give offence. (For example, in the Sikh religion, it is offensive to include anyone playing the part of the Guru in role-play; in the Muslim religion it is offensive to depict human or animal form in the mosque).

Festivals of light

Topic
Celebrations

Resources
- *Lights for Gita* by Rachna Gilmore (Second Story Press)
- Non-fiction books on celebrations
- Examples of artefacts that support sources of light, such as candle stick with candle, diya (or diva) with candle or oil, menorah
- Matches

⚠ Close supervision when using candles and matches.

Keep matches on your person at all times.

Learning objectives
- To learn about similarities and differences among families, communities and traditions
- To share and discuss the practices, resources, celebrations and experiences of various cultures and faiths
- To find out about and identify different sources of light by asking questions

What to do
Circle time
- Show the artefacts and let the children handle them. Ask them to describe the artefacts and what they are made of.
- Light the candles. Ask the children what they notice.
- Read *Lights for Gita*. Ask how Gita was celebrating Diwali.

Extensions/variations
- Read religious stories that relate to the use of light.
- Have lots of picture books showing different festivals.

Science notes
- Darkness is the absence of light.

Links to home
- Ask for the loan of artefacts.

Personal, Social and Emotional Development with Understanding the World and Mathematics

Happy Chinese New Year

Learning objectives
- To talk about own home and community life and find out about other children's experiences
- To understand similarities and differences among cultures, communities and traditions, by sharing and discussing practices, resources and experiences
- To create movement in response to music

Preparation
- Obtain information and pictures on the Chinese New Year and the lion dance from the Internet.
- Find out the date of the Chinese New Year.
- Check access to web page.

What to do
Circle time
- Read *Lion Dancer: Ernie Wan's Chinese New Year*.
- Ask the children if they know which night in Britain is celebrated by lighting fireworks? Refer to the 5th November in the UK.
- The Chinese New Year is also celebrated with fireworks. Ask the children if they know what people do to celebrate the New Year in the UK. (Firework displays are held in many cities and the London display is on television over midnight.)
- Some children may be able to talk about what happens in their countries if they return to families abroad for the holiday.
- Other cultures have their New Year celebrations at different times throughout the year. Each Chinese New Year takes the name of one of the 12 animals. The Chinese New Year celebrations include several rituals which the children can role-play.

Book corner
- Read books about China.
Role-playing
At Chinese New Year:
- The home is thoroughly cleaned. The children can be involved in a clean and tidy of the setting.
- A negative word should not be spoken. Play a game in which the children try not to say 'No!'
- People should wear red (a happy, lucky colour). The children can wear red clothes on

Topic
Celebrations

Resources
- *Lion Dancer: Ernie Wan's Chinese New Year* by Kate Waters and Madeline Slovenz-Low (Scholastic)
- *The Fancy Dress Party Lai Chun and Lai Kit* by Gillian Klein and Simon Willby (Methuens Resources for Reading) – book corner
- *C is for China* by Sungwan So (Frances Lincoln) – book corner
- Brush and pan
- Sign 'DO NOT USE'
- Coloured tissue paper
- Glue and spreaders
- Scissors
- Red card
- Crayons
- Access www.crayola.com for dragon card
- Type in 'Chinese New Year' to access web pages

 Supervise use of scissors.

the day that they perform the lion dance.
- Cutting tools are not used. Do not use knives or scissors on the day of the lion dance.

Extensions/variations
- Make greetings cards with red paper.
- Make paper lanterns.

Links to home
- Ask a Chinese person to advise and help with the preparations.
- Ask parents to dress their children in red clothes for the performance of the lion dance.

This is me

Topic
Myself

Resources
- *Eyes, Nose, Fingers and Toes* by Judy Hindley (Walker)
- *All Kinds of People* by Emma Damon (Tango)
- *All about Me* by Debbie MacKinnon (Frances Lincoln)
- Large hand mirror with handle
- Plastic mirrors
- Paints
- Sticky coloured paper for eyes
- Wool for hair
- Scissors
- Red or pink crayons
- White paper plates
- Glue
- Labels
- Collection of pictures of people of all ages and nationalities
- Practitioner's passport
- Camera

⚠ Don't use glass mirrors.

Learning objectives
- To look closely at similarities and differences between people
- To know some of the things that contribute to a unique child
- To describe self in positive terms

Preparation
- Mix colours of paint for faces – pink, light brown, darker brown. Cut out circles of sticky coloured paper for eyes – blue, brown, pale brown, green.
- Set up a picture display showing people of all ages and nationalities.

What to do
Circle time
- Read *Eyes, Nose, Fingers and Toes* or one of the other books suggested. Ask if everyone looks the same. Discuss whether everyone has the the same colour skin and hair.
- Invite a mixed group of children to talk about their skin colours, hair and eyes. Talk about the children's facial features, naming parts of the face.
- Pass the mirror round the circle and encourage each child to say the colour of their eyes, hair and skin, and how long their hair is.

Art activity
- Support the children making self-portraits using the paper plates. Help them to choose paint colour, hair colour and length, eye colour, etc. Use red or pink crayons for mouths.

Display area
- Support language when looking at the display of faces. Challenge any racist remarks.

Extensions/variations
- Show your passport and explain the importance of passports to identify people.
- Make passports. The children could use a camera to take photographs of each other.

Links to home
- Explain activity. Ask parents to talk about what people look like in terms of colours of skin, hair, eyes.

Related activity
- What a picture! (see page 130)

Who's that knocking at my door?

Learning objectives
● To show interest in different occupations and ways of life
● To know about differences between themselves and others and among families, communities and traditions
● To develop narratives and explanations by connecting ideas or events

What to do
Circle time
● Read the *Postman Pat* story and discuss it with the children. Ask how often the post delivery person calls at their homes. Ask if the post is also delivered to the setting. Encourage them to think of other people who might call at their homes or at the setting.
● Show items, and ask the children to say what they are and who might use them. Ask if the dentist might come to their house. Ask whether other people might, and under what circumstances (for example, meter readers, people to mend broken appliances, etc).
● Talk about safety and remind children that they should never answer the door without an adult.
● Read *ABC I Can Be* and talk about the jobs people do. Support the learning of new vocabulary and developing language skills by introducing the names of various jobs and professions and the equipment people use.

Extensions/variations
● Arrange for a caring professional, such as a police officer or a nurse, to visit the setting and talk about how they can help children.
● Put clothes and items to represent people who help us in the dressing-up box.
● Set up a house in the role-play area and play with children, acting out scenarios involving different people knocking at the door.

Links to home
● Suggest that parents talk with their children about different occupations.

Related activities
● A letter to Santa (see page 139)
● I want to be … (see page 27)

Topic
People who help us

Resources
■ Any *Postman Pat* story by John Cunliffe (Hodder Children's Books)
■ *ABC I Can Be* by Verna Allette Wilkins (Tamarind Books)
■ Non-fiction books showing different careers
■ Items that can be associated with a person's job, for example: letter and parcel = postman; bucket and chamois leather or wiper = window cleaner; newspaper = newsagent or newspaper delivery person; stethoscope = doctor, etc.

⚠ Talk to the children about stranger danger.

Older and older

Resources
- *Grandfather and I* by Helen E Buckley and Jan Ormerod (Puffin)
- *See How You Grow* by Dr Patricia Pearse (Frances Lincoln)
- Telephone(s)
- Pictures of people of different ages
- Labels: Babies, Children, Teenagers, Adults, Elderly people

Learning objectives
- To make observations and talk about similarities and differences, growth and changes
- To become aware that changes occur over a period of time
- To show an interest in the lives of people familiar to them
- To find out and talk about past and present events in own lives and lives of family members
- To use everyday language related to time, size and age

What to do
Circle time
- Read and show pictures from the story *Grandfather and I*.
- Ask what differences the children can see between the young child and the Grandfather (look especially at their heads and faces).

Ask them to explain why the Grandfather does not hurry. You could ask how old they think the Grandfather is. Ask them about their own grandparents.
- Support the learning of new vocabulary and developing language skills by introducing words such as: *today, years ago, now, then, tomorrow, yesterday, older, younger, before, after, when?* and *one day.*

Role-play
- Encourage children to pretend to telephone their grandparent(s) to tell them what they have been doing at the setting. Suggest that they invite them to tea, or organize a trip out with them.

Table activity
- Support the sorting activity. Sort the pictures of people into piles of: babies, children, teenagers, adults and elderly people.

Extensions/variations
- Invite an elderly person into the setting to answer questions about when he/she was a child.
- Ask them to make particular reference to the toys and equipment that were available to them and the lack of much of the technology that we now rely on.
- Sort the pictures into peer groups, then move them around to create an assortment of different types of families.

Links to home
- Ask for volunteers to talk to the children. It would be especially interesting to have a mother and grandmother to visit.
- Try to ensure a good mix of male and female volunteers and a range of ages and generations.

Personal, Social and Emotional Development with Understanding the World and Mathematics

© Mavis Brown and Rebecca Taylor and Brilliant Publications

When I grow up

Learning objectives
● To show interest in different occupations and ways of life
● To show interest in the lives of familiar people
● To make observations and look closely at similarities and differences
● To use everyday language to discuss time and describe past, present and future

What to do
Circle time
● Share the picture book *What Am I?* with the children.
● Ask the children if they can name the jobs the people are doing and whether any of their parents have the same jobs.
● Talk about the clothes the people in the pictures are wearing and about uniforms in general.
● Hold up a hat and ask what job the person does when they wear the hat. The child who answers correctly wears the hat.

Table activity
● Pass round the book *What Am I?* and ask the children what job they would like to do when they grow up.
● Ask if they can say why and support them as they explain.

Role-playing
● Encourage the children to dress up as a person who helps us.
● Encourage the children to act out helping people, eg firefighter putting out a fire, nurse looking after teddy.

Topic
People who help us

Resources
■ *What Am I?* by Debbie MacKinnon and Anthea Sieveking (Frances Lincoln)
■ Pictures of people wearing clothes that identify their job, eg police officer, firefighter, post delivery person, etc
■ Dressing-up clothes, including hats

Extension/variation
● Children who have dressed up could show the rest of the class and say what their job is.

Links to home
● Ask parents if they could talk to their child about their job, and whether they have to wear a uniform or special clothes. Suggest that the children might be able to bring in a photograph to show to the group.

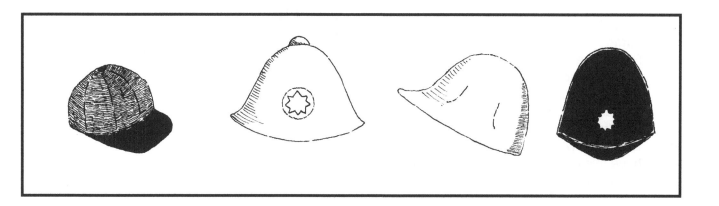

© Mavis Brown and Rebecca Taylor
and Brilliant Publications

**Personal, Social and Emotional Development
with Understanding the World and Mathematics**

Let me entertain you!

Topic
Homes

Resources
- Portable CD player and CD
- Portable cassette player and tape cassette
- Vinyl record and record player
- Television
- Video player with music tape
- DVD player with DVD disc
- 78 vinyl record
- Wind-up record player
- Music box
- Instrument, preferably piano
- Music hall songs

⚠ Tell the children that they must never touch electrical appliances unless an adult gives them permission.

Learning objectives
- To find out and talk about past and present events in own lives and lives of familiar people
- To show an interest in technological equipment and learn how to operate simple items
- To sing familiar songs, learn others and begin to build a repertoire

What to do
Circle time
- Play some music using the video, DVD, tape cassette and CD in turns and encourage the children to look for and comment upon similarities and differences in style and quality.
- Explain how people years ago did not have this technology and discuss how they entertained themselves.

- Show the children a tape cassette and a vinyl record and compare them with a CD.
- Play music box, instrument or wind-up record player.
- Talk about quality of sound.
- Talk about safety with electricity.

Extensions/variations
- Invite children to sing a song, tell a joke or say a rhyme individually.
- Arrange for a practitioner or visitor to play and sing music hall songs at the setting.

Links to home
- Ask parents and grandparents if they could play the piano and sing for and with the children.

Personal, Social and Emotional Development with Understanding the World and Mathematics

Transport in days gone by

Learning objectives
- To begin to differentiate between past and present
- To make observations
- To look closely at similarities and differences, patterns and change
- To become aware that these changes occur over a period of time
- To use everyday language relating to time

What to do
Circle time
- Read *Gumdrop at Sea,* which is about an old car. Ask the children in what way Gumdrop looks older than the other cars. Ask them why the old car has broken down. Point out the old weatherboarded houses and the old fishing boat.
- Support the learning of new vocabulary and developing language skills by introducing *words such as: old, new, modern, vintage* and *classic.*

Collection table
- Point out the time span between some of the models – hundreds or tens of years.
- Ask if the children can pick out an old model and a modern design of the same type of vehicle. Ask for differences they see.
- Ask which vehicles would go faster.

Extension/variation
- Read a *Thomas the Tank Engine* story. Ask what makes Thomas go (coal/steam) and what makes trains go nowadays (electricity, diesel). Talk about safety and keeping away from railway lines.

Links to home
- If any parent or relative has an old or classic car, ask if they would bring it to the setting to show the children.

Topic
Transport and travel

Resources
- *Gumdrop at Sea* by Val Biro (Puffin)
- *Thomas the Tank Engine* by the Reverend W.V. Awdrey (Heinemann Young Books)
- Books for practitioner: *Life in Victorian Times*; *Travel and Transport*, *Travelling through Time*, *Past and Present*, all by Neil Morris (Belitha Press)
- Pictures and toy models of old and modern vehicles

Where do your relatives live?

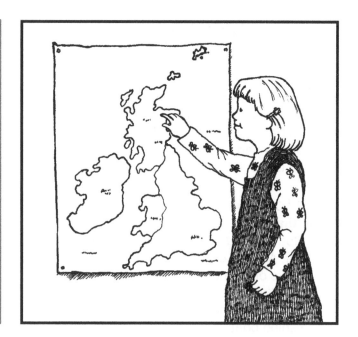

Topic
Families

Resources
- *A Gift for Gita* by Rachna Gilmore (Mantra Publishing) in dual language
- Atlas suitable for age group; outline map of the British Isles
- Outline map of the world (if required)
- Coloured thread
- Name labels that the children have written

Learning objectives
- To show an interest in the lives of familiar people and where they live
- To comment and ask questions about the immediate environment and the natural world
- To share experiences of other countries and cultures, as well as other parts of Britain

What to do
Circle time
- Read *A Gift for Gita,* which refers to Gita's relatives in India. Ask the children where their relatives live, and whether they have visited them.
- Show them on a map of Britain or the map of the world where their relatives live. Fix their name labels to the maps with string. Ask them if they had to travel far to see them, or whether they live near.

- Support the learning of new vocabulary and developing language skills by introducing words such as: *near, far* and *long way.*

Extension/variation
- Invite children to talk about visits that they make to their relatives.

Links to home
- Ask parents to say where their relatives live, that the children have visited or might visit one day.

Personal, Social and Emotional Development with Understanding the World and Mathematics

A day to remember

Learning objectives
- To know about similarities and differences among families, communities and traditions
- To recognize and describe special times or events for family or friends
- To remember and talk about significant events in own experience

What to do
Circle time
- Show photograph of children as bridesmaids/pageboys. Ask if anyone has attended a wedding. Ask about their experiences. Talk about the significance of the occasion. Ask about special clothes, special foods, presents being given and why.
- Read the story *Maisie Middleton at the Wedding*.
- Allow the children to talk about anything funny that happened at any weddings they attended.

Extension/variation
- Involve all children who would like to take part in dressing up and re-enacting weddings.
- Ensure that you do not state that weddings always have to involve a boy and a girl.

Links to home
- Ask for wedding photographs showing their child as a bridesmaid/pageboy.
- Ask parents to talk to their child about their or any relative's wedding if appropriate.

Topic
Celebrations

Resources
- *Maisie Middleton at the Wedding* by Nita Sowter (Picture Lions)
- Dressing-up clothes suitable for wedding role-play
- Wedding photographs

- Ask if parents are prepared to lend bridesmaids' dresses, pageboy outfits, waistcoats and jackets for grooms and white dresses for brides or could help by making accessories.
- Invite parents to watch wedding role-play.

Related activities
- Wrapping paper for a present (see page 125)
- What a picture! (see page 130)

Our baby

Topic
Families

Resources
- *Billy and Belle* by Sarah Garland (Puffin) or
- *New Born* by Kathy Henderson (Frances Lincoln) or
- *Little Brother and the Cough* by Hiawyn Oram (Frances Lincoln)
- *Angel Mae* by Shirley Hughes (Walker)
- For role-play area, set up as a nursery: dolls, baby equipment
- For water play: plastic dolls, towels, flannels, soap

⚠ Supervise water tray.

Learning objectives
- To recognize and describe special times or events for family or friends
- To remember and talk about significant events in own experience
- To express own thoughts and feelings through discussion, role-play and stories

What to do
Circle time
- Discuss siblings with the children and find out who has a younger brother or sister. Ask if they can remember their mummy going to hospital to have the baby.
- Ask if anyone's mummy is expecting a baby now.
- Read *Billy and Belle*, showing the pictures and pointing out the conversations.

- Initiate discussion with the children by asking the following questions: When their mummy had to go to the hospital, where did the children go? Did Billy think about the baby? What did Billy say to his mummy when he left home to go to school? How did you feel when you had a new baby in your home?

Role-play
- Support the learning of new vocabulary and developing language skills by joining in with role-play and chatting as the children play with the dolls as babies.
- Encourage careful carrying, dressing the dolls to go out in the push chair or undressing to go to bed.

Book corner
- Look at the alternative books.

Water tray
- Supervise and support language development at the water tray as the children play at bathing the dolls.
- Ask the children why they must keep the doll's head above the water. Ask them why they must keep the soap out of the doll's eyes.
- Encourage them to dry the doll carefully.

Extensions/variations
- Invite a parent to come into the setting who has a new baby (mother or father) to feed and bathe the baby.
- If it is near Christmas time you could read the book *Angel Mae*.

Links to home
- Invite a parent who is expecting a baby to make regular visits to the setting over a period of time to talk to the children about the changes taking place.
- At Christmas, invite parents to offer help with any aspects of a nativity play.

Related activities
- Who is that? (see page 133)
- All grown up (see page 76)

Burning brightly

Learning objectives
● To celebrate diversity by understanding that different cultures and faiths exist within the community
● To know about similarities and differences among families, communities and traditions and to learn about a festival for one of them
● To use simple tools and techniques competently, safely and appropriately
● To select appropriate resources and adapt work where necessary

Preparation
● Prepare a wooden menorah to show how it can be made.

What to do
Circle time
● Invite visitors into the setting to talk to the children about Hanukkah, Diwali and other festivals from particular religions.
● Read an appropriate version of the story of the victory of the Maccabees over the Syrians.
● Show a menorah (a nine-branch candelabrum).

Woodwork
● Support the children in cutting lengths of 2, 6, 10, 14 and 18 cm wood of 25 mm wide by 5 mm thick. Use a bench hook.
● Stack the wood on top of each other in diminishing lengths, and place by measuring, or by eye, to allow steps (2 cm). Pencil mark along the edge of the lengths.
● Drill a hole in the centre of the 2 cm length, and at the ends of the other lengths to take cake candles.
● Glue the lengths in place or/and nail together. Clamp with G-clamp until set.

Topic
Celebrations

Resources
■ Simple version of the story of Hanakkah (the victory of the Maccabees over the Syrians) – several available on the Internet
■ Menorah
■ Wooden menorah made by practitioner
■ Bench hook
■ Junior hacksaw
■ Small hammer
■ Panel pins shorter than 25 mm if required (blunt before use to prevent wood splitting)
■ Centre punch
■ Wood glue
■ G-clamp
■ At least 50 cm length of planed wood 25 mm x 5 mm
■ Hand drill
■ Drill bit same diameter as candles
■ 9 cake candles

 Do not light the candles. Supervise the use of woodwork tools closely.

Links to home
● Visitor to explain the rituals involved in their religion.

**Personal, Social and Emotional Development
with Understanding the World and Mathematics**

Where do you come from?

Topic
Myself

Resources
- CD player
- Story books and CDs with recorded stories – folk tales from different cultures
- Bilingual story books, eg from Mantra Publishing
- Person able to read dual-language book
- Large world map

Learning objectives
- To compare and learn about similarities and differences among families, communities, cultures and traditions
- To share experiences of living in different countries around the world and observing a variety of home cultures
- To experience and compare a variety of stories and languages from around the world

What to do
Circle time
- Choose those countries from which the children in the setting originate, to begin the discussion. Find the counties on the map. Then talk about other countries and language(s) spoken.
- Read a folk story from another country without showing the pictures at first. Ask the children if they can guess in which country the story is set. Thinking about whether it is a hot country or a cold country, and what kinds of animals appear in it, may offer clues.

CD player
- Show the children how to use the CD player so they can listen to the stories.

Extension/variation
- Visitor (parent) or practitioner to read dual-language book to children.

Links to home
- Arrange for parents who can speak another language to take part in activities and to teach the children some words, phrases, rhymes and songs in their first languages.
- Ask parents to talk to their children about their cultural heritage.

Related activity
- Where do your relatives live? (see page 86)

Personal, Social and Emotional Development with Understanding the World and Mathematics

Help me to cross the road

Learning objectives

- To show interest in familiar people and different occupations
- To talk about past and present experiences in own lives
- To recognize that a range of technology is used in familiar equipment for everyday life
- To engage in imaginative role-play based on own first-hand experiences
- To recognize the school crossing patrol person

What to do

Outdoor activity

- Play at crossing the road. Support the learning of new vocabulary and developing language skills by introducing words such as: *stop, go, wait, carefully, left, right, cross, look, listen, think* and *concentrate*. Teach the children how to use the Green Cross Code, but ensure that they understand that they still must not cross the road without an adult while they are too small for car drivers to see them easily.

Table activity

- Support use of construction kits (eg Lego®) to make wheeled vehicles, road ways, etc.

Floor activity

- Support play on play mat with model vehicles.

Visit

- Organize a trip out of the setting to visit the High Street and use the pelican crossing, and observe the traffic lights.

Extension/variation

- Ask the school crossing patrol person to come and talk to the children.

Links to home

- Ask parents to reinforce the use of pelican crossings, and the Green Cross Code. Ask volunteers to come and help to practise the Green Cross Code in the playground.
- Ask parents to take their child to use a pelican crossing when visiting the High Street.

Topic

People who help us

Resources

- Outdoor wheeled toys
- Road drawn on the playground
- Wooden traffic lights
- Construction kit to make wheeled vehicles, road ways, traffic lights, etc, eg Lego®
- Play mat with town streets
- Model vehicles

 Check local authority and setting guidelines for out of setting visits.

Personal, Social and Emotional Development with Understanding the World and Mathematics

Victorian homes

Topic
Homes

Resources
- Photographs of children living in the Victorian era and Victorian homes
- Books for practitioner: *Life in Victorian Times*, *Home and School* and *Work and Industry* by Neil Morris (Belitha Press)
- Information from the Internet (practitioner accesses)
- CD-ROM *Treasure Chest* (Granada Learning)
- Victorian lace cloth, clothes, woven wool
- For role-play area: sepia prints on the wall and lace tablecloth over a table, a large green plant and peg rug
- Galvanized bucket
- Galvanized bath

⚠ Supervise children closely around galvanized bath and bucket and when using needles for sewing.

Learning objectives
- To show interest in different occupations and ways of life from history
- To develop an understanding of changes over time
- To use everyday language related to time

What to do
Circle time
- Show photographs of people and children in the Victorian era and their homes. Ask the children how different the Victorian homes looked in comparison with their own home.
- Girls had to do a lot of sewing and help around the house, and all the children had to work in the garden, if they had one.
- Show a lace tablecloth, woven wool and Victorian hand-sewn clothes. Explain that some children even worked in factories that made cloth, and some worked underground.
- Explain that poor people did not have water taps in their home. They had to go down the road to a tap, and collect all their water in a bucket. To have hot water, it had to be heated on a fire, where all their cooking was done. There was not a bathroom. They had to have a bath in front of the fire.
- Ask the children to comment upon their own bathrooms and kitchens.

Role-play
- Support turning the role-play area into a Victorian sitting room.
- Encourage the children to sew or read in the area.

Extensions/variations
- Support children in filling up the galvanized bath with water using a bucket.
- Talk about how long it takes, how heavy the bucket of water is and the quantity (volume) of the water.

Science notes
- Use the word volume, not capacity.

Links to home
- Ask if parents or grandparents could lend any of the items in the resources list, or borrow from museums and find in craft shops.

My home

Learning objectives
- To talk about own home and community life and find out about other children's experiences
- To know about similarities and differences between types of homes for families and within communities
- To describe homes, parts of buildings and features liked and disliked
- To express opinions on the built environment

What to do
Circle time
- Show the photographs of homes for the children to pick out the type of home in which they live.
- Ask the children to describe the outside of their home and to say what they like or dislike about it.
- Ask if they have a garden. Ask what the road outside is like (busy or quiet?).
- Support the learning of new vocabulary and developing language skills by introducing words such as: *flat, maisonette, bungalow, house, farm, house-boat, semi-detached, detached, terraced, caravan, mobile home, trailer, city, town, countryside, street, road, avenue, lane, cul-de-sac* and *estate*. Also supply words to describe parts of buildings and building materials, such as: *roof, windows, doors, walls, brick, stone, pebble dash, wood, tile* and *slate*.

Small world activity
- Support play mat activities. Ask the children if the road is busy and where the nearest bus stop, shop, school is.
- Ask what kind of homes are in their street/road.
- Ask them to build single and double storeyed homes and blocks of flats with the bricks.

Extension/variation
- Read *Moving House*. Ask the children if anyone remembers moving home. Talk about moving home and what has to be planned.

Links to home
- Talk with parents about what types of homes they have, in order to make the topic relevant to individual children.

Topic
Myself

Resources
- Photographs of homes in different environments (city, town, countryside, suburban), and of different styles (flat, maisonette, bungalow, semi-detached, detached, terraced) and different ages (timber-framed, bay window of 1930s, Georgian, modern)
- Play mat of a town
- Vehicles
- Small world people
- Small bricks to represent buildings
- *Moving House* by Anne Civardi and Stephen Cartwright (Usborne Publishing Ltd)

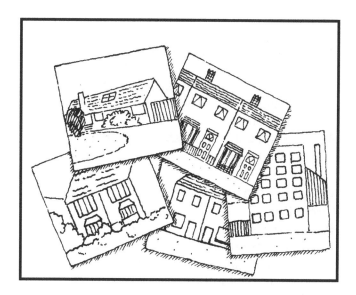

- Ask parents to talk about different homes with their children and to come into the setting to talk if they have a particularly interesting home.

Personal, Social and Emotional Development with Understanding the World and Mathematics

Where do you work?

Topic
People who help us

Resources
- Non-fiction picture books and pictures showing places and people working, eg building site, farmer, bus driver, school and teacher, office and secretary
- Dressing-up clothes
- For sand play: small digging vehicles
- For water play any items associated with a job, eg boats for fishermen, nets (eg from vegetables, fruit)
- Building site toys
- Construction: Lego® to make vehicles

⚠ Check local authority and setting guidelines for visits outside your setting.

Learning objectives
- To know about similarities and differences between people and their jobs, and among families, communities and traditions
- To show interest in different occupations and ways of life
- To find out about features in the local built environment

What to do
Circle time
- Show and discuss non-fiction books and pictures showing people working.
- Ask what job the people are doing and where they work.

Sand play
- Ask the children if they are excavating to build a house or factory.
- Ask if it is a good idea to play on a building site.

Water play
- Ask if the fishermen caught any fish. Ask if the fishermen got cold and wet in the storm?

Dressing up
- Suggest the children select dressing-up clothes for a person in a particular occupation.

Construction toys
- Ask the children to make a vehicle that belongs on a building site.

Extension/variation
- Take the children to visit an interesting place of work, such as a fire station or a building site, but talk about the dangers that can be present.

Links to home
- Ask parents to talk to their children about the jobs they do, and, if possible, to take them to visit their workplace (even if it is just to see the outside of the building).
- Ask whether any parents could arrange for the children to visit their workplaces.
- Invite parents to accompany practitioners and children on trips out of the setting to visit workplaces.
- Be sensitive to any families who are out of work and suggest that children could talk to any family member or friend about their workplace.
- Ensure that children understand that being at home to look after the family is also a very important job.

Related activity
- I want to be … (see page 27)

Personal, Social and Emotional Development with Understanding the World and Mathematics

My painted house

Learning objectives
- To compare and learn about similarities and differences among families, communities, cultures and traditions
- To name and talk about features and materials used in the built environment and look closely at patterns
- To represent own ideas and create appropriate designs through art and craft work

Preparation
- Prepare designs.

What to do
Circle time
- Read *My Painted House, My Friendly Chicken and Me,* showing pictures. Ask the children if their house is painted. Discuss the material from which their house is made. Compare the African house to their own home. Compare their lives with the African children's life and family.
- Support and encourage children to use descriptive language and to name colours and building materials, such as: *concrete, pebbledash, corrugated iron, straw, mud, stone, brick, slate, tiles, wood, plastic* and *glass.*

Art activity
- Show examples of Ndebele designs painted on houses.
- Encourage the children to make their own designs by gluing strips of coloured paper or gummed paper geometric shapes on to off-white or black paper.

DVD
- Share with the children a DVD showing African village life.

Extensions/variations
- A Ndebele village can be created from decorated boxes with roofs and small world farm animals.
- Visit a cultural centre or museum that contains ethnic art.

Links to home
- Ask parents to talk to their child about their culture and the structure of their home.

Topic
Homes

Resources
- *My Painted House, My Friendly Chicken and Me* by Maya Angelou and Margaret Courtney-Clarke (Clarkson Potter Inc./ Publishers) or
- *Chidi Only Likes Blue* by Ifeoma Onyefulu (Frances Lincoln)
- Small boxes
- Off-white and black paper
- Glue
- Scissors
- Large copies of Ndebele designs (available on Internet)
- Coloured strips of paper
- Gummed paper geometric shapes
- Small world farm animals
- DVD of African village life
- Children's hymns

 Supervise use of scissors.
Check local authority and setting guidelines for outside visits.

Other children's toys

Topic
Toys

Resources
- Non-fiction books showing toys from other cultures
- Book for practitioner *Children Just Like Me* (Dorling Kindersley)
- Toys from other cultures (borrow from museum or parents)
- World map
- Labels and thread

What to do
Circle time
- Pass round appropriate toy(s) and ask the children to make observations.
- Show the pictures of toys and talk about them. Ask the children what is different about the toys from other cultures compared with their similar toys (eg it could be a difference in material).
- Ask from where they think the toys were obtained. They might have been bought from a shop or made in a factory, or parents might have made them themselves.
- Discuss how the chidlren would play with the toys.
- Some of them might be smaller grown-ups' things, like tools. The children might have toys like this in the setting, especially in the construction and role-play areas.
- Point out and mark countries on the world map with labels and thread.

Links to home
- Ask if any parents have toys from other countries to show the children.

Learning objectives
- To compare and learn about similarities and differences between self and others and among families, communities, cultures and traditions
- To know that other children don't always enjoy the same things and to show sensitivity to others' needs and feelings
- To experience and compare a variety of toys from around the world

Personal, Social and Emotional Development with Understanding the World and Mathematics

Understanding the World

The World

- Children develop their understanding of the world around them from birth, as they watch people, animals, objects and features, listen to voices and sounds, absorb smells and explore with hands, feet and mouths. As they grow, they remember and recognize familiar people and objects and learn what things are like, how they work and what they are used for.

- Practitioners should encourage children to play with small-world models that represent different environments, such as farms, garages, railways, streets or houses. These 'small worlds' may be compared with the local area and children encouraged to notice details and features in their own environment, allowing them to become aware of similarities and differences, patterns, sequences and changes.

- It is important to take children outside into the garden and on walks to the park and other places of interest, so that they can ask questions and talk about what they observe in the natural world. They should be introduced to plants, animals, mini-beasts, natural and found objects and will want to have conversations about why things happen and how things work.

- Parents may be able to take their children on visits to farms, zoos and wildlife parks, on journeys by car, bus and train and to specific environments, both natural and man-made, such as beaches, forests, stations, airports and shopping centres. First hand experiences provide ideas to stimulate children's imaginative play and a deeper understanding of the world around them. They will often choose to recreate their own experiences through play over and over again, until new interests replace them.

- Experiments, cookery and early science activities, using malleable materials, food ingredients, magnets, sand, mud, water and natural, found, recycled and craft materials, will allow children to explore the properties of objects and media through all their senses. They should gradually develop an understanding of growth and decay, recognize changes that occur over periods of time and learn how to be careful and caring towards living things and to protect the environment.

- Children's questions may be answered, interests extended and other experiences introduced through sharing books, encyclopedias and dictionaries, DVDs and CD-Roms, computer games and programs and information from the internet with their families at home and with practitioners within early years settings.

What does it feel like?

Topic
Myself

Resources
■ *Goldilocks and the Three Bears* and *The Princess and the Pea* (Ladybird do nice editions)
■ *Your Senses* by Angela Royston (Frances Lincoln)
■ Feely boxes or bags – have objects which are: hard, soft, scratchy/ rough (eg sandpaper), prickly (eg bristle brush), cold (eg metal), etc

 Do not put anything breakable inside the feely boxes or bags.

Learning objectives
● To investigate objects and materials by using the sense of touch
● To know about similarities and differences in relation to objects and materials
● To express ideas effectively, showing awareness of listeners' needs

Preparation
● Prepare feely boxes for class discussion. Make holes in boxes for hands and arms to go through, place a piece of fabric across the hole so the child cannot look in.

What to do
Circle time
● Read the story *Goldilocks and the Three Bears* and highlight the feel of the beds and ask if the chairs would feel hard or soft.
● Alternatively, read *The Princess and the Pea* and ask why the princess could feel the pea.

● Prompt them by asking *What does it feels like? Can you guess what it is?* The child might answer, for example, *It's scratchy, it feels like a hedgehog, it is a brush.*
● Support the learning of new vocabulary and developing language skills by introducing descriptive words such as: *hard, soft, prickly, fluffy, warm, cold, smooth, rough, stiff* and *floppy.*

Extension/variation
● Create a display with story picture books with feely pages.

Science notes
● Nerve endings can detect heat, pain and pressure.

Links to home
● Encourage parents to talk about what things feel like when handling toys and objects with their children.
● Suggest that parents allow their children to help to put away the laundry and the shopping.

Personal, Social and Emotional Development with Understanding the World and Mathematics

Balancing sandy shapes

Learning objectives

- To look closely at similarities and differences between materials
- To explain why some things occur and talk about changes
- To construct with materials, observe patterns and experiment with design, form and function
- To observe that some shapes can balance on top of each other and be used for building

What to do

Circle time

- Read *The Three Little Pigs* and talk about the materials from which the houses were made.

Sand tray

- Ask the children to build a house from sand using the different shaped moulds. Talk about how successful each shape was.
- Ask which material was better to use, the wet or dry sand.

Science notes

- The properties of sand are changed when water is added. The activity examines how some shapes when built up into a wall are more stable than others.

Links to home

- Ask parents to point out the pattern of bricks in a wall.

Topic

Shapes

Resources

- *The Three Little Pigs* (Ladybird do a nice edition)
- Moulds to make shapes of bricks, spheres, stars, pyramids, animals, vehicles, etc.
- Spades, spoons and scoops
- Buckets
- Wet and dry sand

Do plants drink water?

Topic
Water

Resources
- Two bunches of flowers (tulips work well)
- Two vases
- Water
- Two clear plastic containers – one containing plain water and the other with red food colouring mixed in the water
- Head of celery with leaves attached
- Large magnifying glasses

⚠ Check on any allergies to food dyes. Discourage children from tasting the celery as good practice in science.

Learning objectives
- To make observations of plants and explain why some things occur and talk about changes
- To show care and concern for a living plant

Preparation
- About three weeks in advance of the activity, divide the sticks of celery with leaves attached into the two containers. One should contain water only, while the other should also contain the food colouring. Leave both of them in the light.

- On the day before the activity, put one bunch of flowers into a vase of water and put the other into a vase without any water.

What to do
Circle time
- Point out that the flowers are drooping in the empty vase. Talk about having pot plants and cut flowers at home. Ask the children why do they think that the flowers in one vase have gone droopy?
- Draw attention to the coloured leaves of the celery in the coloured water. Encourage children to record what they see by making drawings. Discuss why the leaves have turned red.

Extension/variation
- Cut the coloured celery stick in half. Use a magnifying glass to investigate.

Science notes
- Cells are mainly composed of water. Without water the cells collapse, and the plants lose their rigidity (go floppy). The coloured dye marks the passage of the water through vessels called xylem up the stem into the leaves, where it is lost by transpiration.
- This experiment will take a few weeks for the red colour to stain the celery leaves.

Links to home
- Ask parents if they could spare a few flowers from their gardens to bring into the setting.

Make a rainbow

Learning objectives
- To make observations of features of the natural world and explain why some things occur
- To choose particular colours to use for a purpose
- To represent ideas through art and design

Preparation
- Choose a sunny day.
- Set up tank near window (see diagram).

What to do
Circle time
- Let the sun shine through the water in the tank. Tilt the mirror inside the tank so that the sun hits the mirror and reflects on to the ceiling. Ask the children what they notice about the rainbow and whether they can name the colours.

Table activity
- Invite children to cut out an arch shape from the centre of a piece of card. A practitioner can place sticky-backed plastic over the arch-shaped hole and temporarily fasten the card to a table, with the sticky surface facing upwards.
- Support the children cutting small strips of cellophane to fit rainbow shape, overlapping colours to create the range of colours of the rainbow. Practitioners can support the children in cutting out the arches when completed and sticking them to windows in the setting.

Extension/variation
Art activity
- Let the children paint rainbows on the blue paper. Let them experiment by mixing two different colours at a time.

Science notes
- The rainbow is part of a circle but we only see half of it. The red is always at the top of the curve and blue is under the curve. To see a rainbow the sun has to be low in the sky (early morning or evening). The sunlight is bent by the raindrops and is split into the separate colours.

Topic
Colours

Resources
- Clear plastic tank with water
- Plastic mirror to fit into tank
- Scrap card as template
- Sticky-backed plastic
- Coloured cellophane: red, yellow, blue or coloured cellophane paper sweet wrappings
- Scissors
- Paint: red, yellow, blue
- Pale blue sugar paper
- Paint brushes

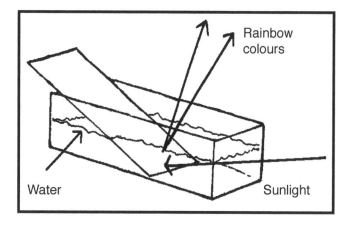

Rainbow colours

Water

Sunlight

Links to home
- Ask parents to encourage their child to name the colours that they see.

Personal, Social and Emotional Development with Understanding the World and Mathematics

Where did the puddle go?

Topic
Water

Resources
- *Alfie's Feet* by Shirley Hughes (Red Fox)
- Bucket of water
- Chalk
- Metre stick
- Camera
- Watering cans
- Water pistols
- Empty washing-up liquid bottles
- Clean decorating paintbrushes
- Buckets of water
- Plastic aprons

Learning objectives
- To make observations and comparisons of features of the immediate environment
- To talk about why things happen and explain some changes that take place
- To comment and ask questions about aspects of the natural world

Preparation
- This activity should be done on a fine sunny day.

What to do
Circle time
- Read *Alfie's Feet*. Ask the children what happens to puddles of water.

Outside activity
- Pour a bucket of water onto a piece of hard ground in the outdoor area to make a puddle.
- Arrange for two children to mark around the outside of the puddle with chalk.
- Take a photograph of the puddle with some indication of its size (line up some children along the length of it).
- After a while, go back outside, chalk around the outside of the puddle again and photograph it. Ask what the children think has happened to the puddle.
- Support the learning of new vocabulary and developing language skills by introducing words relating to time (*yesterday, today, tomorrow, last week, next week, soon, days*), and words relating to size (*big, little, small, bigger, smaller, huge, tiny*).

Outdoor water play (if a hot day)
- If appropriate, let the children play with the squirty water toys when they are wearing swimming costumes. Be sensitive to different religious dress codes.
- Comment on how the splashes of water are drying up.

Extension/variation
- Provide water and brushes for the children to paint outside. After a suitable time go outside. Ask what they think has happened to their paintings.

Links to home
- Warn parents that the children will be playing with water and make sure that all children have enough spare clothes.

Seeing clearly

Learning objectives

● To look closely at similarities and differences in relation to objects and materials
● To talk about why things happen and how things work

What to do

Circle time

● Ask what sort of things the children can see through, and what sorts of material they cannot see through.
● Play a version of *I spy*, 'I can see through this and it begins with the sound ... .'

Collection display

● Draw attention to the collection of materials and make suggestions for experimentation.
● Support the learning of new vocabulary and developing language skills by introducing the words: *opaque, transparent* and *translucent.* Explain the meanings of the words.

Science notes

● Materials that let light through are transparent. Opaque materials block light and will cast a shadow. Translucent materials let some light through.

Topic
Colours

Resources
■ Collection of transparent, translucent and opaque materials – to include coloured glass, and greasy paper (which was used in windows before glass was cheap enough for everyone)
■ Magnifying glass
■ Spectacles
■ Sunglasses

⚠ Supervise children with glass materials.

Links to home
● Suggest to parents that they point out all the things that their child can see through.

Personal, Social and Emotional Development with Understanding the World and Mathematics

Let's get mixing

Topic
Food and shopping

Resources
- *Finished Being Four* by Verna Allette Wilkins (Tamarind Limited)
- Wet area: newspapers; plastic aprons; label pots/jugs with name of solid/liquid and a number or letter
- Plenty of clear plastic beakers and spoons
- Solids: cornflour, flour, sugar, salt, pepper, dried herbs, bicarbonate of soda, jelly crystals (**Note**: use cornflour instead of wheat flour, then children with a wheat allergy can do this experiment)
- Liquids: water, milk, white vinegar, lemon juice, lemonade, washing-up liquid, cooking oil, golden syrup, blackcurrant drink

⚠ Check if any children have wheat or other food allergies.

Learning objectives
- To investigate different solids and liquids by using their senses
- To look closely at similarities, differences, patterns and change when mixing substances
- To talk about why things happen and how things work

What to do
Circle time
- Read *Finished Being Four*. Ask what was being mixed together.

Table activity
- Explain to the children that they are going to see what happens when they mix things together. Encourage them to mix together one solid with one or two liquids. Show the children how to investigate in a methodical manner.
- Talk about their observations using appropriate everyday language that the children can understand, asking questions such as: *What does it smell like? What do you see? Can you pour it? Does it dissolve/ sink/float? Has it changed colour?*

Extension/variation
- Try mixing some solids with cooking oil, golden syrup, or blackcurrant drink.

Science notes
- Vinegar and bicarbonate of soda will fizz with carbon dioxide bubbles. The children might compare bubbles in lemonade with the bubbles of washing-up liquid. Salt, sugar and jelly crystals dissolve, but the jelly also melts (turns to liquid) when heated. Lemon juice will curdle milk.

Links to home
- Provide a list of the chosen solids and liquids and invite parents to donate any of them.

**Personal, Social and Emotional Development
with Understanding the World and Mathematics**

Beans means ...

Learning objectives

- To develop an understanding of growth, decay and changes over time
- To observe and talk about the features of a seed and a plant
- To comment on the natural world and explain why some things occur

Preparation

- Soak the beans before the activity.

What to do

Circle time

- Read *Jody's Beans*. Pass around some whole beans and some that are split open to show the embryo plant inside. Explain that the seed has its own food to help it to grow.
- Support the learning of new vocabulary and developing language skills by introducing the names of the parts of the plant, such as: *seed, skin, root, shoot, leaves* and *stem*.

Table activity

- Set up a growing jar (see drawing). Help the children to put a cylinder of kitchen towel into the beaker. Put newspaper in the centre so that the kitchen towel is pushed against the beaker. Put the bean between the side of the beaker and the kitchen towel. Pour water into the middle.
- Look at the bean every day to see growth. You could take photographs. Encourage the children to describe the changes.

Extension/variation

- Children could sequence drawings of a germinating bean.

Science notes

- When opening the bean, the children should see that a skin with a scar covers the bean, and inside there are two halves and a small embryo plant can be seen between them.
- Oxygen but not light is needed for germination. Food is not needed as the seed has its own food store.

Topic

Gardening

Resources

- *Jody's Beans* by Malachy Doyle (Walker)
- One broad bean or runner bean for each child, plus a few spare
- Clear plastic beakers
- Sheets of kitchen roll
- Newspaper
- Water
- Labels
- Drawings of the stages in germination

 Wash hands after handling the materials.

Links to home

- Inform the parents about the activity. Ask them to let their child help in the garden or to water houseplants.

Who dropped that?

Topic
Health

Resources
- *Tidy Up Titch* by Pat Hutchins (Red Fox)
- Bag of objects to include natural materials, man-made materials and recyclable materials (**Caution**: ensure no sharp edges)
- Two boxes for sorted items
- 'Litter' such as empty wrappings
- A3 card
- PVA glue and spreaders

⚠ Stop children from picking anything off the ground. Wash hands after handling soil or anything on the ground. Check local authority and setting guidelines for outside visits.

Learning objectives
- To know about similarities and differences in relation to natural and man-made objects and materials
- To begin to understand which materials may be recycled
- To make observations of the immediate environment and talk about changes
- To show care and concern for the environment and to express opinions, commenting upon features that they like or dislike.

What to do
Outside activity
- Take the children to see a dustcart collecting rubbish and/or take the children out of the setting and point out any litter and ask what should be done with it.

Circle time
- Encourage the children to voice their opinions on the environment near to the setting, and where they live.
- Examine some of the contents of the bag and discuss which materials are natural, and which are man-made.

Table activity
- Ask the children to sort the contents of the bag into natural and man-made. Support language development.

Extensions/variations
- Read *Tidy Up Titch*. Involve the children in tidying up the room, and afterwards ask what changes they have noticed.
- Make a poster with samples of litter glued to it.

Science notes
- Steel, unlike aluminium, is magnetic, so it can be separated from aluminium drinks cans.

Links to home
- Send a letter home to parents for permission to take their children for a walk out of the setting. Ask for volunteer helpers to join you on the walk.
- If their household recycles their rubbish, encourage parents to take child to the recycling bins.

Related activity
- Attractive toys (see page 122)

Personal, Social and Emotional Development with Understanding the World and Mathematics

Mirror, mirror, on the wall

Learning objectives
- To comment and ask questions about familiar objects and materials
- To talk about why things happen and how things work

What to do
Circle time
- Read the story book *Rosie's Room* and discuss the different reflections. Ask the children if they can think of any other things in which they can see their reflection.

Table activity
- Encourage children to investigate the mirrors and other shiny objects in your setting. Draw attention to and make suggestions for experimentation.
- Support the learning of new vocabulary and developing language skills by introducing words such as: *reflect, reflection, shiny, light, same, identical* and *mirror image*.

Science notes
- If two mirrors are placed at right angles facing each other, and a small toy put between the mirrors, the image reflects back and forth to make many images. Metals are reflective if shiny. Crumpled aluminium foil reflects the image into different directions so the image is broken up.

Links to home
- Suggest to parents that they point out everything that gives a reflection on the way home, eg in a dark shop window, in a puddle, in a car wing mirror.

Related activity
- Make a rainbow (see page 101)

Topic
Homes

Resources
- *Rosie's Room* by Mandy and Ness (Milet Publishing Ltd)
- Reflecting surfaces such as plastic mirrors, shiny metal objects, smooth aluminium foil, shiny spoons, dark washing-up bowl with water, shiny hologram paper, kaleidoscope
- Small coloured plastic toys

 Supervise children with scissors and glass and other fragile materials.

Do not use glass mirrors.

The power of water

Topic
Water

Resources
- Water tray
- To make a water wheel: plastic rectangles cut from milk containers or fizzy drinks bottles, corks, cotton reels, straws, dowelling that can fit through centre of cotton reel
- Scissors
- Glue and glue spreaders
- Sticky tape
- String, Wool
- Stapler
- Toy windmill
- Squeezy bottles
- Plastic bottles
- Washing-up liquid bottles (empty)
- Plastic screw top bottles with 3 small holes punched into the side
- Plastic syringes, tubing and funnels
- Sink with tap

 Supervise use of scissors.

Learning objectives
- To talk about why things happen and how things work
- To look closely at similarities, differences, patterns and changes while using a variety of materials and resources
- To select appropriate resources and adapt work when necessary
- To select materials, tools and techniques to construct with a purpose in mind and achieve a planned effect

What to do
Water play
- Suggest ways for the children to try to stop the water from coming out of the bottle that has holes in the side.
- Try putting the screw top on and cover the holes one or two at a time.

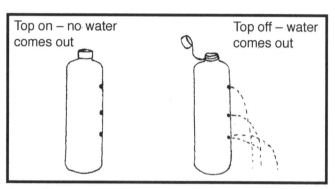

Top on – no water comes out

Top off – water comes out

- Note where the water comes out of the three holes.
- Use water in a squeezy bottle to push corks along.
- Show how water finds its own level with the plastic tubing and funnels attached.

Extensions/variations
- Show windmill. Discuss wind power.
- Make a water wheel. Compare with windmill.

Discuss whether fins should be flat or curved.

Test under running tap to determine which design turns fastest.

Science notes
- Pressure of water increases with depth of water.
- Water pressure experiment works because of atmospheric air pressure.

Links to home
- Invite parents to come into the setting to help and support the children making water wheels and water pressure toys.

Good morning

Learning objectives
● To comment and ask questions about aspects of the natural world
● To talk about why things happen
● To develop an understanding of differences and season changes that occur over a period of time

Preparation
● This activity should be done on a sunny day.

What to do

Circle time
● Read *What is the Sun?* If it is winter, ask if it is light when the children get up in the morning. If it is summer and the sun is shining, go outside to look at shadows.

Outdoor activity
● Put a tall stick into a container of sand. With some chalk, draw the shadow line that the stick makes.
● Go back an hour later and mark the shadow again. Talk about why the shadow has moved. (CARE: Tell the children not to look directly at the sun.)
● Support the learning of new vocabulary and developing language skills by introducing words such as: *light, dark, bright, shine, daytime, nighttime, summer, winter* and *shadows*.

Extension/variation
● In summer, encourage children to make shadow shapes

Links to home
● Ask parents to talk about the shadows that the sun makes.

Topic
Seasons

Resources
■ *What is the Sun?* Reeve Lidbergh and Stephen Lambert (Walker)
■ Containers of sand and tall stick to cast a shadow
■ Chalk

Where can we play?

Topic
Toys

Resources
- *Miffy at the Playground* by Dick Bruna (Egmont Books)
- Balls
- Bicycles
- Small world people, playground equipment

⚠ Check that the playground has soft landing surfaces. If going to an adventure playground ensure that you have sufficient adult support. Check local authority and setting guidelines for visits outside your setting.

Learning objectives
- To comment and ask questions about aspects of the familiar and the natural world
- To find out about features in the local natural and built environments and how environments might vary from each other
- To know the importance for good health of physical exercise and talk about ways to keep healthy and safe
- To show understanding of the need for safety and consider and manage some risks

What to do
Circle time
- Read *Miffy at the Playground*. Ask the children what kind of equipment they would see in a playground and which kinds of toys they could play with in a playground.
- Ask if it would be safe to play with a ball/ bicycle inside. Ask for reasons for their answer.

- Ask the children if it would be safe to play with a ball outside by the road. Discuss road safety.
- Ask if they think it would be safe to play on their own at the park or playground. Discuss the matter.
- Ask the children if there are any other places where they should not play.

Outside activity
- Play with balls or bicycles outside. Talk about how much space there is, and where they are in relation to the perimeter fence and gate.

Visit
- Go to an adventure playground. Talk about the equipment and how the children use it.

Extension/variation
Small world activity
- Encourage the children to set up a playground for their toys to play in. Ask the children to describe how their toys are playing with the equipment.

Links to home
- Ask parents to take their child to a park or playground.
- Ask parents for permission to take their child to visit an adventure playground.
- Invite parents to accompany practitioners and children on trips out of the setting to visit parks and playgrounds.

Related activity
- Faster and faster (see page 119)

Personal, Social and Emotional Development with Understanding the World and Mathematics

Mini-beast hunt

Learning objectives
- To identify and make observations of mini-beasts
- To look closely at similarities and differences in relation to living things
- To show care and concern for living things and their environment
- To look closely at similarities and differences between mini-beasts
- To show care when handling a small animal

Preparation
- Lay upturned plant pots, large stones, logs of wood, compost in piles over an area of soil a few months before you intend to carry out the activity.

What to do
Circle time
- Read *Jasper's Beanstalk*. Talk about the story.

Outside activity
- Take the children out to the garden to look for mini-beasts. Draw attention to where they live, how they live, how many legs, or if they have wings.

Extensions/variations
- Find pictures of mini-beasts in non-fiction books.
- Look closely at any mini-beasts brought back to the setting. Afterwards release them back into the garden.

Science notes
- Mini-beasts are invertebrates. Those with legs are arthropods and their skeleton is on the outside. Insects have six legs and three parts to their body; spiders have eight legs and two parts to their body; crustaceans have ten legs; and centipedes and millipedes have many legs.

Links to home
- Ask parents to look out for spiders' webs with their children.

Topic
Gardening

Resources
- *Jasper's Beanstalk* by Nick Butterworth and Mick Inkpen (Hodder Children's Books)
- Non-fiction books showing invertebrates
- Magnifying glasses
- Pooter
- Jars for collection of mini-beasts

⚠ Check children have been vaccinated against tetanus. Supervise children using magnifying glasses. Wash hands after working outside.

Boil an egg

Topic
Water

Resources
- One egg for each child, plus two extra
- White saucer
- Large pan and water
- Stove
- Three egg timers with ringing bells
- Cutting board
- Knife
- Spoon
- Crayons
- Drawings showing sequence of how to boil egg

⚠ Keep children away from the boiling water completely. Before giving the hard boiled eggs to the children, put them in cold water to cool right down. Although the children are not going to eat the eggs, encourage good hygiene.

Learning objectives
- To make observations of eggs and talk about their features
- To explain why some things occur and talk about changes

Preparation
- This is an ideal activity to do before Easter.
- Cook the eggs away from the classroom.

- Hard boil one egg per child plus one extra. (Leave one egg uncooked.)
- Prepare some drawings to illustrate the sequence of boiling an egg.

What to do
Circle time
- Crack open a raw egg onto the saucer to see what it is like at the start. Cut open the hard boiled egg for comparison. Ask for descriptions and comparisons.
- Ask how you boil an egg. Use the drawings to help sequence.
Art activity
- Support decorating hard boiled eggs.

Extension/variation
- Some towns have egg rolling races at Easter. Have an egg rolling competition down a grassy slope.

Science notes
- Hot water can be used to cook food. The temperature stays constant at boiling point. Protein goes hard when heated; The white of the egg is protein and water. The yellow yolk is the fat store for the chick. If you look closely on the yolk you will see a tiny white dot, which is the egg that will turn into the chick if fertilized.

Links to home
- Ask parents to talk with their children about the appearance of boiling water and reinforce the safety rules whenever they are near hot water.

Personal, Social and Emotional Development with Understanding the World and Mathematics

Overflowing with water

Learning objectives
● To make observations of objects and materials and talk about why some things occur
● To look closely at patterns and change by noticing how solids displace volumes of liquids
● To use everyday language to talk about and compare size, weight and capacity

What to do
Circle time
● The story *Mr Archimedes' Bath* involves the concept of the displacement of water. Read the story with the children.
● Ask the children to predict what will happen at each stage of the story. Use toys that will sink to demonstrate the story.

Water tray
● Supervise water tray and support the learning of new vocabulary and developing language skills by introducing words such as: *overflow, space, size, float, sink, more* and *less.*

Extensions/variations
● Make a hole in a clear plastic container near the top, so water coming out can be caught in a measuring jug.
● Fill up the container to the hole.
● Let the children investigate how much water comes out when a toy is put into the container (see diagram).

Science notes
● A submerged object displaces its own volume. When an object is placed into water this volume of water is pushed out of the way. By putting a toy into a container with a hole in the side at the top, you can catch the water and measure it. The child can compare the amount of water collected for a small object and a large object.

Links to home
● Suggest to parents that they talk about how the level of water goes up when their children get into the bath. They could

Topic
Water

Resources
■ *Mr Archimedes' Bath* by Pamela Allen (Puffin)
■ Water tray
■ Toys that sink
■ Containers for water large enough to take toys, pebbles, plastic teddy bears
■ One clear plastic container in a washing-up bowl with small toys that sink
■ Measuring jug or beaker
■ Jug of water
■ Cloth for mopping up spills

⚠ Supervise water tray.

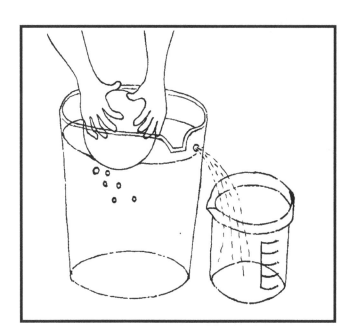

consider making a mark on the bath before a child gets in and watching to see how the level increases.

Icebergs

Topic
Water

Resources
- Ice blocks made inside small yoghurt pots
- Water tray
- Ice lolly moulds
- Diluted fruit juice
- Access to freezer

⚠️ Do not let children put the ice in their mouth, as it can cause ice burns. Make sure that the frozen lolly is not too cold before giving it to the children.

- Point out that the ice is sticking up over the top of the small yoghurt pots. Suggest finding out if the ice, yoghurt pot full of water, and yoghurt pot with ice all float or sink.
- Ask the children to listen carefully and say if they hear anything when the ice goes into the warm water (it should make cracking noises).
- Talk about how slippery the ice is.

Extension/variation
- Make fruit juice ice-lollies.

Science notes
- Water turns solid (ice) at temperatures at and below 0ºC. When water turns to ice, it expands and so ice is less dense than water, and it floats on top of water.

Links to home
- Ask parents to talk to their child about things that are frozen, and what happens when they get warmer.

Learning objectives
- To make observations of ice and its properties
- To explain why some things occur and talk about changes

Preparation
- Freeze water in small yoghurt pots.

What to do
Water tray
- Put the empty pots, the frozen blocks of ice and ice still inside their yoghurt pots into the water.
- Support the learning of new vocabulary and developing language skills by introducing words such as: *cold, warm, freeze, frozen, melting, smaller, float, sink, crack* and *slippery.*

The power of air

Learning objectives
- To talk about why things happen and how things work
- To look closely at similarities, differences, patterns and changes while using a variety of materials and resources
- To select appropriate resources and adapt work when necessary
- To select materials, tools and techniques to construct with a purpose in mind and achieve a planned effect

What to do
Circle time
- Read the story *My Best Friend* and initiate a discussion with the children, inviting them to contribute their feelings about the wind.
- Ask what they would use to measure how strong the wind was, eg feather for breeze, rolling drinks can in high wind.

Outside activity
- Try out the children's suggestions to measure the wind.
- Try sticking together layers of newspaper, or different sizes of sheets of newspaper.
- Discuss which method would be more accurate.

Extensions/variations
Craft activity
- Make a carp streamer out of tissue paper. Support the learning of new vocabulary and developing language skills by introducing words such as: *making, glue, stick, cut, punch, decorate, tube, streamer* and the names of shapes.
- Decorate a rectangle of tissue paper with coloured tissue shapes.
- Roll it into a tube and stick down the edge.
- Fly the streamer on a windy day, or tie to a tree. On Boys' Day (Japan) carp streamers are fixed to trees.

Links to home
- Invite parents to come into the setting to help and support the children in making carp streamers.

Related activity
- Flying high (see page 123)

Topic
Weather

Resources
- *My Best Friend* by Pat Hutchins (Red Fox)
- *Air is All Around You* by Franklyn M. Branley (Harper Collins Children's Books) – alternative title
- Coloured tissue paper
- Parcel tape
- PVA glue and spreaders
- Stapler
- Strong cotton or nylon thread (CARE)
- White card
- Crayons (bright colours and black)
- Aluminium foil
- Scissors
- Hole punch
- Feathers
- Newspaper sheets
- Drinks can
- Twigs

⚠ Keep away from overhead wires. Avoid string getting caught around child's neck.

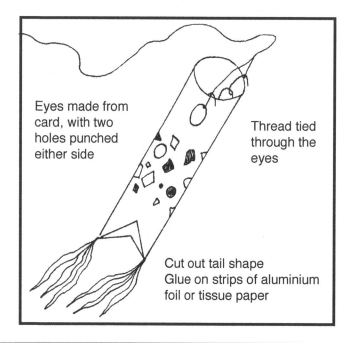

Eyes made from card, with two holes punched either side

Thread tied through the eyes

Cut out tail shape
Glue on strips of aluminium foil or tissue paper

Personal, Social and Emotional Development with Understanding the World and Mathematics

Changes

Topic
Animals

Resources
- *From Tadpole to Frog* by Wendy Pfeffer (Harper Collins Children's Books)
- Non-fiction books showing life cycles of frog, eg *Guess What I'll Be* by Anni Axworthy (Walker)
- Frog spawn in tank with lid
- Calendar
- Pictures of developing tadpole for calendar
- Pictures of life cycle of frog for sequencing
- Strips of sugar paper
- Pencils, scissors, glue

 Do not allow children to put their hands into the water, as it may be contaminated with dangerous bacteria. After use, wear gloves and use disinfectant to clean the tank, away from children.

Learning objectives
- To make observations of animals and talk about changes
- To develop an understanding of similarities and differences, growth and changes over time
- To use talk to organize, sequence and clarify thinking, ideas and events
- To develop narratives and explanations by connecting ideas and events
- To communicate meaning, using letters, words or other marks and pictures

Preparation
- Collect the frogspawn, using rubber gloves.
- Life cycle pictures of the frog

What to do
Circle time
- Show non-fiction books and talk about the changes in the life cycle of the frog.

- Draw attention to the developing legs of the tadpoles. Look at differences and change between the young and the adult frog
- Encourage and support children in talking about what is happening, introducing new words and descriptions as appropriate to stimulate language development.

On-going activity
- Show the children the frog spawn. Discuss with the children how you should look after the developing tadpoles (eg feed them and keep their water clean).
- Encourage the children to observe the tadpoles every day. Support recording development of the tadpoles on a calendar.
- Maintain the tank until almost turned into frogs and then release back to the local pond.

Extension/variation
- Give the children sequence pictures of the life cycle of a frog and help them to write their own commentary. Make it into a zigzag folding book.

Science notes
- Sexual reproduction begins with a fertilized egg. The young of frogs and toads are totally different from the adults. This process of change is called metamorphosis. Amphibians do not have scales, but bare skin.

Related activities
- Mini-beast hunt (see page 111)

Personal, Social and Emotional Development with Understanding the World and Mathematics

Growing plants

Learning objectives
● To observe, identify and talk about some features of growing plants.
● To develop an understanding of growth, decay and changes over time
● To explain why some things occur and talk about changes
● To examine and make observations about soil
● To investigate the ideal conditions for plant growth

What to do
Table activity
● Support the learning of new vocabulary and developing language skills by introducing names of parts of the plant, such as: *stalk, leaf, bud, stem, petal, flower* and *root.*
● Supervise potting the plants.
● Ask if anyone has a greenhouse at home. Discuss why plants grow well in the greenhouse.
● Set up some plant pots in the sun and some others in the shade. Invite children to take it in turns to water both sets of plants.
● Measure the plants once a week. Ask the children to suggest how they should measure the growth.

Extensions/variations
Table activity
● Support the children as they put water and soil into plastic jars, put lids on tightly (supervise) and shake them. If they then leave the jars to settle, they will see what happens to the mixture. Ask if they know why it happens.
● Investigate the effects of light, warmth and soil (food). Some plants can be put in the refrigerator, some in a dark cupboard, and some can be planted without compost.

Outside activity
● Plant bedding plants in the garden or in tubs outside.

Topic
Gardening

Resources
■ Plant pots
■ Compost
■ Large trays
■ Newspaper to reduce mess
■ Plants already with growth eg tomatoes or flowering plant, but not fully grown
■ Water
■ Jug
■ Wall chart for results
■ Paper strips to measure height
■ Camera
■ Plastic jars with screw-top lids, quarter filled with soil (not compost)

 Wash hands after working with soil or compost.

Science notes
● Certain conditions are needed for growth (light, water, soil/food, warmth). Soil contains humus, which is the source of food for the plant. The soil will separate out into stones at the bottom, fine silt, and plant material at the top.

Links to home
● Talk to parents about the gardening activity and suggest that they might be able to give their children a small piece of garden, tub or windowbox in which to grow plants.

Bouncing balls

Topic
Toys

Resources
■ Balls of different sizes, materials and colours
■ Metre stick

⚠ Timid children may not want to take part. Make sure there is room to bounce the balls so that no-one is hit by them.
Keep away from perimeter fence (road or residential gardens).

Learning objectives
● To observe and compare objects, recognizing similarities, differences and patterns
● To talk about why things happen and how things work
● To measure and compare using standard and non-standard units
● To safely use and explore objects and materials, experimenting with form and function

What to do
Outside activity
● Offer the children a selection of balls to bounce. Before the investigation ask which ball the children think is going to bounce the best.

● Drop one ball at a time. Try each ball at least three times. Notice how high the ball goes. and agree on which one bounced the best.
● Encourage the children to record their findings on a chart. Talk about whether they predicted the right ball and why they thought that ball would be the bounciest.
● Help the children to record their findings on a chart. Did they predict the right ball? Why did they think that ball was the bounciest?

Extension/variation
● Older children could work collaboratively, using metre sticks. Ask if their test was fair. (Were the balls dropped from the same height and on to the same surface?)

Science notes
● Children will often identify colour as a variable. They should compare the results from different sized balls and different materials separately.

Links to home
● Ask parents to help their children to look at home for any other toys that bounce.

Faster and faster

Learning objectives

- To observe and compare objects and materials, recognizing similarities and differences
- To look closely at patterns and change by noticing how variables affect each other
- To talk about why things happen and how things work
- To use everyday language to compare sizes, speeds, gradients and surfaces
- To explore characteristics of objects and materials and use mathematical language to describe them

What to do

Circle time

- Talk about getting somewhere faster. Read *Wheels*. Ask which vehicles in the story had wheels.
- Ask what can happen if a vehicle goes too fast. Talk about road safety.

Construction

- Ask the children which they think would make a vehicle go faster, having big wheels or small wheels. Investigate.
- Support the learning of new vocabulary and developing language skills by introducing correct names for the parts of vehicles.

Extensions/variations

- Roll cars down ramps/slopes to find out which car travels furthest and/or fastest.
- Using constructed cars, find out if they will break apart if they crash.

Links to home

- Ask parents to point out to their children the speed limit signs for roads.

Topic
Transport and travel

Resources
- *Wheels* by Shirley Hughes (Walker)
- Small cars and other wheeled toys
- Construction kit suitable for making wheeled vehicles
- Hardboard ramps
- Equal sized building blocks

⚠ Take care where the children are playing on the floor so the cars do not cause a hazard.

Where do animals live?

Topic
Animals

Resources
- Non-fiction books on the habitats of animals, eg *Looking at Animals* series by Moira Butterfield (Belitha Press); *Where Am I?* by Moira Butterfield (Belitha Press); *Who Are You?* and *In the Sea* by Vic Parker and Ross Collins (Franklin Watts)
- *Nine Ducks Nine* by Sarah Hayes (Walker) – shows the countryside
- *The Picnic* by Ruth Brown (Andersen Press) – shows animals that live underground
- *One Snowy Night* by Nick Butterworth (Picture Lions) – shows woodland animals
- For water play: plastic ducks, fish, frogs, shells, stones, seaweed, etc.

⚠ Check local authoruty and setting guidelines for out of setting visits. Supervise water play.

Learning objectives
- To find out about and identify features of animals' habitats within the natural world
- To recognize and name a variety of animals and make observations of them
- To show care and concern for living things and the environment

What to do
Circle time
- Read one of the suggested books and show the pictures, asking the children to name the animal.
- Talk about what the habitat looks like.
- Talk about how the animal is adapted to its environment.

Book corner
- Support reading the suggested books.

Water play
- Support play by asking which animals would be living together in a pond/in the sea.
- Encourage the children to talk about the selected animals.

Extension/variation
- Organize a visit to an environmental area.

Links to home
- Invite parents to accompany practitioners and children on a visit to an environmental area.

Related activities
- Mini-beast hunt (see page 111)

Where does rain come from?

Learning objectives
● To comment and ask questions about aspects of the familiar and the natural world
● To make observations, explain why some things occur and talk about changes
● To find out about features in the local environment and the natural world

What to do
Circle time
● Read the folk tale *The Story of Running Water*. Ask the children whether they think it was a true story. Tell them that rain falls and runs down the mountain to the rivers, lakes and then to the sea. The water evaporates into clouds. The rain falls from the clouds.
● Support the learning of new vocabulary and developing language skills by introducing words such as: *flow, pour, gush, trickle, drip, liquid, sea, river, lake* and *pond.*

Sand tray
● Support use of sand tray to focus the idea of rain falling on a mountain, and the water running down it.
● Pile up dry sand, then pour water over it.
● Watch the water soak in then run down the side and out from the bottom of the sand.

Table activity
● Support setting up dissolving salt in water, then leaving it to evaporate. Also set up tap water at the same time.
● Pour tap water and salty (sea) water on to saucers and leave uncovered on a shelf. Label the saucers.
● After a week, ask the children where the water has gone?
● Support the learning of new vocabulary and developing language skills by introducing words such as: *dissolve, evaporate* and *vapour.*

Topic
Water

Resources
■ *The Story of Running Water* by Joanna Troughton (Cambridge University Press)
■ Non-fiction books about water, rivers and the sea
■ Salty water
■ Tap water
■ Dark coloured plastic saucers
■ Labels
■ Watering cans, sieves
■ Jug of water
■ Dry sand

Extension/variation
● Place a tin filled with ice on the table and the children can see the condensation form on the outside of the tin.

Science notes
● Salt will be left after the water evaporates.
● If the tap water is hard, white residue will be left behind.

Links to home
● Suggest that the children try the experiment with sugar instead of salt at home.

Attractive toys

Topic
Toys

Resources
- Selection of toys that contain magnets
- Steel paper clips
- Plastic paper clips
- Electromagnet (see diagram)
- A few bar magnets
- A few small steel objects that can be lifted up by the bar magnets (to be hidden in the sand tray)

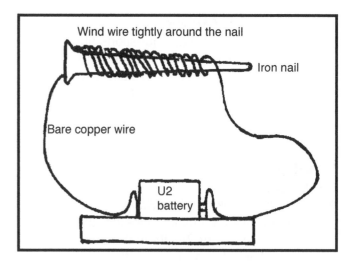

Wind wire tightly around the nail

Iron nail

Bare copper wire

U2 battery

⚠ Take care that items hidden in the sand have no sharp edges.
Paper clips should not be used as buried treasure in case a child bends one open, and puts it back into the sand.

Learning objectives
- To observe and compare objects and materials, recognizing similarities and differences
- To look closely at patterns and change by noticing how materials affect each other
- To talk about why things happen and how things work

Preparation
- Bury small steel objects in the sand tray.
- Prepare and check electromagnet.

What to do
Table activity
- Invite the children to investigate the electromagnet (see diagram). Supervise them closely. Ask them to try to pick up steel paper clips and plastic paper clips. Ask if they can guess why a magnet can't pick up the plastic ones.

- Investigate toys containing magnets and talk about how they work.
Sand tray
- Ask the children if they can find the 'hidden treasure' in the sand tray without digging. Give them a variety of bar magnets with which to experiment.

Extension/variation
- Make a moving magnetic toy by fastening a steel paper clip to a 2D paper model. Use the bar magnet to make the model move.

Science notes
- Only steel, iron, cobalt and nickel are attracted to magnets. Other metals are not magnetic.
- Electric motors contain magnets, but this investigation does not include this concept. As electricity can magnetize iron (or steel), a simple electromagnet for picking up paper clips can be made.

Links to home
- Invite parents to donate unwanted fridge magnets for the children to play with.
- Could parents show their child the magnetic seal on the fridge door, and/or any other magnetic catches on cupboards?

Flying high

Learning objectives
- To comment and ask questions about aspects of the familiar world, including aeroplanes and other things that fly
- To talk about why things happen and how things work
- To look closely at similarities and differences and understand how shape and streamlining affects flight
- To use and explore materials and techniques, experimenting with design, form and function

Preparation
- Collection table of model aeroplanes.

What to do
Circle time
- Invite the children to handle the model aeroplanes and look at pictures of aeroplanes. Ask what they notice about their shape.
Outside activity
- Support folding different designs of paper aeroplanes.
- After testing them, talk about why they think certain designs flew the furthest or stayed longest in the air.

Extension/variation
- Make aeroplane shapes using construction kits. Make a model airport.

Science notes
- The shape of the wing and streamlining will affect the flight of the paper dart. Adding Blu-tac® to the nose sometimes helps the balance of the dart.

Links to home
- Ask parents to talk to their child about travelling on an aeroplane.

Topic
Transport and travel

Resources
- Non-fiction books about aeroplanes
- Scrap paper suitable for folding
- Blu-tac®
- Model aeroplanes
- Construction kit

⚠ Take care when children are throwing their darts. Keep away from perimeter fences.

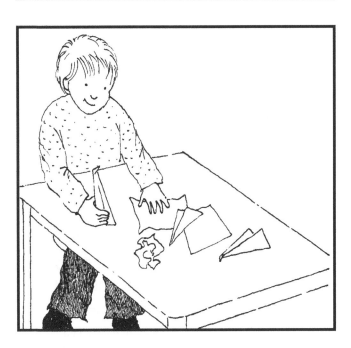

Understanding the World

Technology

- Children now grow up with technology all around them and adults must provide opportunities for them to explore, experience and use it appropriately.

- Toys with knobs to turn, levers to press, flaps to lift and compartments to open stimulate young children's curiosity, manipulative skills and interest in mechanisms. They may learn to operate wind-up toys, programmable toys and those with remote controls.

- Practitioners should support children as they acquire basic skills in operating ICT equipment, such as a computer, a keyboard and a mouse, and other resources, such as a camera, a CD player and a microwave.

- Children should understand that a range of technology is used in their homes, nurseries and schools and also in offices, factories and other public places. They need to learn to identify technology and to select and use it appropriately, both independently and with adult support. Practitioners may demonstrate the use of a laminator, a photocopier or a cooker during an activity. Children may learn to complete simple programs on a computer, play CDs or take photographs without help.

- By the end of the foundation stage, most children in a group setting or school will be able to use a mouse or a keyboard to interact with software designed especially for the early years, play simple games on a computer, operate programmable and mechanical toys and tell practitioners when they think that technological equipment should be used. At home, they may know how to use remote controls to operate the television and DVD player and how to turn on and play with a games console.

- These skills ensure that children feel confident in their use of technology and can continue to develop and refine their knowledge and understanding of ICT and machines as they grow older.

Personal, Social and Emotional Development with Understanding the World and Mathematics

Wrapping paper for a present Technology

Learning objectives
- To select and use technology for particular purposes
- To select tools, techniques and materials to achieve a planned effect
- To use a simple program on a computer to make a repeating pattern

Preparation
- Access suitable drawing program on the computer.

What to do
Circle time
- Read *Alfie Gives a Hand* stressing the wrapping up of the present. Ask when we receive presents. Check calendar for birthdays

Design activity
- Look at *Fun with Patterns*.
- Ask the children to make their own wrapping paper. Discuss, compare and describe the commercial wrapping papers, noting the repeating patterns and shapes.
- Support the learning of new vocabulary and developing language skills by introducing the names of shapes, materials and colours. Encourage children to discuss and plan their designs.
- Teach skills to make a potato print or print using shapes, or toys.
- Allow the children to choose the method of printing.
- Closely supervise cutting the potato.

Extensions/variations
- Use wrapping paper to wrap up present.
- Design and print wallpaper for room.
- Use computer to make repeating pattern.
- Initiate play in the role-play area involving shopping for presents.

Links to home
- Ask for scraps of wrapping paper and wallpaper with repeating patterns.

Topic
Shapes

Resources
- *Alfie Gives a Hand* by Shirley Hughes (Red Fox)
- *Fun with Patterns* by Peter Patilla (Belitha Press)
- Potatoes
- Spoon
- Knife
- Cutting board
- Small plastic (washable) toys
- Plastic shapes
- Paints
- Flat tray with flat cleaning sponge inside for paint for printing
- Examples of wrapping paper
- Calendar showing birthdays of the children in the class
- For home corner: toys, paper for wrapping, register and money
- Computer with drawing program

Quack

Topic
Animals

Resources
- *Quack! Said the Billy Goat* by Charles Causley (Walker)
- CD-ROM *Maisy's Playhouse* (TDK)
- CD-ROM *Fun on the Farm with Barney* (Microsoft) (*see Hootin' Annie's Barn*) alternative title
- CD-ROM *Amazing Animals* (Dorling Kindersley)

Learning objectives
- To select and use technology for particular purposes
- To complete a simple program on a computer
- To play cooperatively, taking turns with others

Preparation
- Load the CD-ROM on the computer and select the appropriate game.

What to do
Circle time
- Read the story *Quack! Said the Billy Goat* and talk to the children to ensure that they know what the correct sounds are for each of the animals.

Computer activity
- Support use of CD-ROM.

Extension/variation
- Look at CD-ROM *Amazing Animals*.

All change

Learning objectives
- To show an interest in ICT
- To recognize that a range of technology is used in familiar equipment for everyday life
- To make observations, explain why some things occur and talk about changes
- To understand the uses of everyday technology

What to do
Collections display
- Make suggestions and draw the children's attention to features of the resources on the collections table.
- Support the learning of new vocabulary and developing language skills by introducing words such as: *light, heat, warmth, shine, reflect, change* and *display*.
- Demonstrate how to use the heat sensitive strip to test their temperature.
- Show how the LCD on the watch face disappears when they wear the Polaroid sunglasses.

Extension/variation
- The children could experiment with crumpled aluminium foil and a torch in the dark.

Science notes
- Chemicals called liquid crystals are used extensively in the electronics industry in displays (such as in a watch) because they are sensitive to small amounts of electricity. Thin layers of these chemicals change colour with different temperatures.
- Light is refracted inside the fine glass optic fibre, and so can be carried along its length.

Topic
Colours

Resources
- Any heat sensitive products, eg Feverscan
- Fibre optic torch or Christmas tree or floral display light with flashing optic fibres
- Wrist watch with liquid crystal display
- Polaroid sunglasses
- Sheet of aluminium foil
- Torch
- Dark den

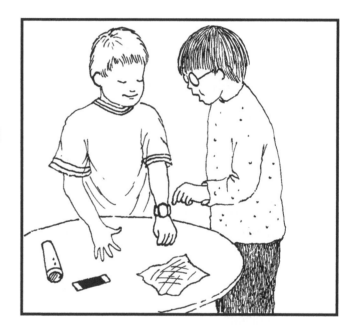

Across space

Topic
Homes

Resources
- Catalogues with pictures of televisions and radios
- Weekly television programme magazines
- Television
- Radio
- A3 sugar paper
- Glue and spreaders
- For collection: slinky, fibre optic torch, pictures of satellites in space, old radio and screwdriver

⚠ Tell the children that they must never touch electrical appliances unless an adult gives them permission.
They must never try to take a television apart as it continues to carry an electrical charge.
Remove electric wire from the radio.

Learning objectives
- To show interest in ICT equipment that receives information, such as television and radio
- To recognize that a range of technology is used in familiar equipment for everyday life in places such as homes, schools and nurseries

What to do
Circle time
- Switch on the television and the radio briefly at different times, to ensure that children understand what they do and the similarities and differences between them.
- Talk about how the television and radio are receiving information from a long way away.
- Talk in general terms about the different types of television reception that are, or have

been, available, including satellite, cable, terrestrial and digital.

Collection table
- Support playing with the slinky (this shows wave motion).
- Take apart the radio. A practitioner needs to remain with this activity at all times, closely supervising children and ensuring that they understand that they must never attempt to do this without adult permission and support.

Extension/variation
- Tear out pictures from the catalogues of televisions, radios and sheets from a weekly television programme magazine to make a collage.

Science notes
- Information can be carried along electromagnetic waves. These waves can be 'bounced' off satellites in space to get round the curvature of the earth.
- Signals for cable television (and telephone) are carried along optic fibres.

Related activity
- All change (see page 127)

Personal, Social and Emotional Development with Understanding the World and Mathematics

Apples in this pie

Learning objectives
- To recognize that technology is used in places such as homes, schools and nurseries
- To select and use technology for particular purposes
- To select the tools and techniques needed to work appropriately with resources
- To look closely at similarities, differences and changes

What to do
Cooking activity
- Pay attention to hygiene.
- Closely supervise cutting apples. Support language development describing smell, appearance, taste and texture of apples.
- Show how to make pastry. Support language development describing the changes in the texture of the ingredients as they are rubbed together.
- Using a 50 gram yoghurt pot allows the children to measure out the ingredients themselves.
- Mix the salt into the flour.
- Rub the fat into the flour until it looks like breadcrumbs.
- Stir in the water, and knead until smooth. Add more flour if the mixture is sticky.
- Roll out the pastry. Cut round the medium sized plate to make a circular base.
- Add chopped up apple to the centre and sprinkle sugar over it.
- Make suggestions on how to wrap the pastry around the apple slices, eg fold the pastry over, cut strips of pastry and lay a lattice over, lay another circle on top, etc. Allow children to choose, and decorate the pastry.
- Place on well greased baking tray.
- Cook in moderate oven (gas mark 5 or electric 375°F/175°C) for 10 to 15 minutes until golden brown.
- Practitioner to put pastries into the oven, and remove them, and supervise clearing up.

Snack time
- After the apple pastries have cooled, let the children taste them and compare with the uncooked apple and pastry.

Topic
Seasons

Resources
- Words for 'Sing a Song of Sixpence'
- Chopping board & knife
- Cooking tray and oven
- Medium sized plate
- Fork
- Tablespoon & teaspoon
- Rolling pin
- Mixing bowl
- Small jug with water
- 50g yoghurt pots
- Greaseproof paper

Ingredients
- 4 yoghurt pots plain flour (= 125g)
- 1 yoghurt pot soft margarine (= 60g)
- 1/4 teaspoon salt (practitioner to add)
- 3/4 full yoghurt pot of cold water (= 50ml)
- 1/4 large apple, peeled & chopped (about 50g)
- 1 heaped teaspoon sugar to sweeten apples
- 1/4 teaspoon ground cinnamon (optional)

 Teach safe use of tools.

Extension/variation
- Teach the children the song 'Sing a Song of Sixpence'.

Links to home
- Invite parents to come into the setting to help with the cookery activity
- Ask parents to contribute ingredients or small sums of money to buy them

What a picture!

Topic
Families

Resources
- *When Willy Went to the Wedding* by Judith Kerr (Collins Picture Lions)
- Wedding photograph album
- Film camera
- Digital camera
- Photograph albums of children's families

Learning objectives
- To explore a camera and show interest in how it works
- To learn how to operate simple equipment
- To understand the uses of everyday technology
- To select and use technology for particular purposes
- To talk about home and community life and share the experiences of others

Preparation
- Check that all of the children have had a previous experience of having their photograph taken. If any families do not do this outside the setting, offer the activity after a professional photographer has been to take an annual group photograph or practitioners have taken pictures of the children engaged in activities for their learning journeys.

What to do
Circle time
- Read *When Willy Went to the Wedding*. Ask at what other occasions families like to take photographs. Show the family albums, asking for comments from the children.

- Open a film camera and show the children how to load a film into it. When the pictures have been taken, show them how to wind the film back and take it out and explain that it must be developed and the prints processed in a pharmacy or shop.
- Explain that a digital camera works differently, storing pictures in its memory without a film, and that the photographs can be loaded into a computer and printed out.
- Show the children how to take photographs with both types of camera and support them while they learn and practise.

Small group activity
- Let the children take photographs of each other.

Extensions/variations
- Use a camera to record activities that are on-going, such as the growth of plants, the construction of a model, or activities that are taking the children several days to complete.
- Take photographs of small groups of children and stick them on a collage, eg as centres of flowers, as passengers on a bus, etc.

Links to home
- Ask parents to bring in a photograph album of their family, including their child. (Ask them to put an identifying label on the album.)
- Ask for a wedding album to show the children.

Related activities
- This is me (see page 80)
- Who is that? (see page 133)
- All grown up (see page 76)

Wibble wobble

Learning objectives
- To recognize that a range of technology is used in familiar equipment for everyday life in places such as homes, schools and nurseries
- To select and use technology for particular purposes
- To learn how to operate simple equipment
- To talk about why things happen and how things work
- To make observations, explain why some things occur and talk about changes

What to do
Cooking activity
- Ask the children if they can think of any way of making water go hard without freezing it. Encourage all ideas.
- Help the children to make up jelly as per instructions. An adult should always place it in the microwave and take it out again. Select a child to press the button to switch it on. When cool, let another child pour the jelly into the mould.
- Make up some squash using the same quantity of water. Put this and jelly into the refrigerator.
- When the jelly is hard, take it and the squash out of the refrigerator. Point out that the jelly has gone hard but the squash has not.

Sand tray
- Pretend to make a cake or jelly, using various moulds.

Water tray
- Give children plastic jugs, small items to stir in and spoons for stirring.
- Ask the children if the items are dissolving.

Extensions/variations
- Eat the jelly. Discuss how it tastes.
- Use jelly block instead of crystal. It will take longer to melt and dissolve.

Science notes
- Jelly molecules form a lattice as it cools down. The microwaves give energy to water, which becomes hot.

Topic
Water

Resources
- To make jelly from crystals: microwave oven, microwaveable jug, sugar-free jelly crystals, water, tablespoon, jelly mould, refrigerator
- Clear bowl, water and squash (same colour as jelly)
- Sand play: making a cake, various moulds
- Water play: spoons for stirring, small items to stir, plastic jugs
- Dishes and spoons for children to eat jelly
- Block of jelly
- Scissors

 Although a microwave oven does not get hot, heat from the liquid is conducted to the container, which does become hot.

Supervise water tray

- Microwaves are part of the electromagnetic spectrum that includes sound, light and radio waves.

Links to home
- Invite parents to donate jelly ingredients or small amounts of money to buy them.

Related activity
- Let's get mixing (see page 104)

Happy festival!

Topic
Celebrations

Resources
- *Celebrations* books (Ginn Reading Books)
- Computer with drawing program
- Colour printer
- Examples of different types of greetings cards
- Glue
- Card
- Crayons
- Access www.crayola.com
- 2 publish – program to make a card from 2 simple Infant Video toolbox. See www.2simplesoftware.com

Learning objectives
- To show interest in and compare different illustrations and print in greetings cards designs
- To understand that pictures and print carry meaning
- To select and use technology for particular purposes
- To manipulate a mouse, coordinate actions, select from a web page and use a drawing program on a computer
- To represent ideas, thoughts and feelings through art and design

Preparation
- Load appropriate drawing program on the computer.
- Access www.crayola.com for card designs.

What to do
Circle time
- Read an appropriate book from *Celebrations*.
- Show greetings cards and ask the children to classify and sort them, eg which Christmas cards have snow, animals, Santa, etc. Ask them to compare Christmas cards with other festival greetings cards.

Computer activity
- Support use of computer, in particular Save and Print.
- Select card design and print. Colour in using computer program or crayons.

Extension/variation
- Using a computer program, design, draw and colour a greeting card.

Links to home
- Ask for donations of unwanted greetings cards for children to sort, compare and classify.

Related activity
- Wrapping paper for a present (see page 125)

Who is that?

Learning objectives
- To select and use technology for particular purposes
- To learn how to operate a piece of simple equipment
- To understand why things happen and how things work
- To recognize familiar people by looking or listening

Preparation
- Record the voices of individual children and practitioners before the activity.
- Write name cards for each child.

What to do
Table activity
- Play a game of 'Who is that?' with the tapes and photographs. The practitioner plays the tape. The children match the photograph with the recording of the voice. If they can, the children could also pick out the name card of the child on the tape.

Extensions/variations
- Encourage children to record themselves and each other singing or saying nursery rhymes, individually or as a group.
- Support children who choose to use the tape recorder to record other sounds around the setting and play them back for others to guess and identify.

Links to home
- Ask parents to bring in photographs of their children, with names marked on the backs.

Related activity
- What a picture! (see page 130)

Topic
Myself

Resources
- Tape recorder
- Blank tape
- Photographs of each child
- Cards

Does it need batteries?

Topic
Toys

Resources
- Different moving toys (including mechanical) which need batteries
- Bicycles with cogs
- Small push-along toys
- Catalogues of programmable toys and toys that need batteries
- Construction kit with cogs and gears
- Tools for dismantling the old toys.

⚠ Do not allow children to put the batteries in their mouths.

Learning objectives
- To show interest in technological toys and equipment
- To develop skills in making toys and equipment work to achieve desired effects
- To recognize, select and use a range of technology
- To talk about why things happen and how things work
- To look closely at similarities, differences, patterns and change

What to do
Collection display
- Investigate moving toys, asking children to think about how the toys move and what gives them their energy.
- Ask what will happen when all the energy in the battery is used up.
- Ask them what they have to do to make the mechanical toys move.

Outside activity
- Investigate how bikes work and what people do to make them move.

Extensions/variations
- Invite the children to dismantle the old toys, under supervision.
- Demonstrate how to take out and replace batteries, pointing out the – and + symbols on the batteries. Supervise very closely if children attempt to do this for themselves.
- Children could make moving toys using cogs and gears.

Science notes
- Energy is the capacity to do work. Different forms of energy are used to move the toys. Electrical energy is useful as it can be changed into sound, light, motion and heat. The chemical energy in the battery is changed into electrical energy.

Links to home
- Ask for donations of old toys that do not work that can be taken to pieces.

Personal, Social and Emotional Development with Understanding the World and Mathematics

Let's find out about animals

Learning objectives
- To select and use technology for particular purposes
- To manipulate a mouse, coordinate actions and find out how to access information from the Internet

Preparation
- Check the setting's system for accessing the Internet.
- Find suitable web pages.

What to do
Computer activity
- This activity will involve a practitioner giving full support at the computer. Ask the children to choose which animal they want to see.
- Type in the name of the animal in the search box, then select the previously determined suitable web page.
- Show the children how they can get back to the list by using the Back button.

Topic
Animals

Resources
- Computer with modem and access to Internet

⚠ Check web pages first to determine their suitability.

Personal, Social and Emotional Development with Understanding the World and Mathematics

Bars and bills

Topic
Food and shopping

Resources
- *Teddybears Go Shopping* by Susanna Gretz (A & C Black)
- Itemized shopping bills
- Tins and boxes with bar codes
- CD-ROM and Playset Playskool Store (Hasbro Interactive)

⚠ Check local authority and setting guidelines for out of setting visits.

Learning objectives
- To recognize and show interest in technology used in familiar places, such as shops
- To discuss and understand the function of some ICT equipment

What to do
Circle time
- Read the story *Teddybears Go Shopping*. Ask the children what the teddy bears had to help them to remember what they wanted to buy.

- Discuss how they could know what they had bought without unpacking their bags (eg by looking at the bill).
- Show the children the bar codes on the packaging and the itemized shopping bills.
- Have a discussion about the function of bar codes.

Computer activity
- Play with the Playset Playskool Store (this is a toy till with a scanner and credit card swiper, and fits over the computer).

Extension/variation
- Take the children to the local supermarket that uses a bar code scanner and watch the procedure at the check-out.

Science notes
- As bar codes identify every product, shops can keep control of their stock. If the customer has a loyalty card, an individual's buying preferences can be determined.

Links to home
- Ask parents to supply clean, empty packaging with bar codes.
- Ask parents to take their child shopping in the supermarket and let them put the food into the shopping trolley.
- Invite parents to accompany practitioners and children on the visit, to ensure a safe 1:2 ratio of adults to children.

Switch on

Learning objectives
- To identify and compare different electrical items, to learn their functions and how to programme them
- To recognize that a range of technology is used in familiar equipment for everyday life in places such as homes, schools and nurseries
- To learn about safety with electricity

Preparation
- Make labels for each of the appliances in the setting. Use Blu-tac® to attach them.

What to do
- Read *Doing the Washing*.
- Show the children pictures from the catalogue and ask them to name the appliances and say what they do.
- If your setting has any of the appliances, let the children look more closely at them.
- Talk about safety with tumble driers, microwaves, ovens, etc, and the importance of keeping water away from electrical appliances.

Extension/variation
- Children could bring in pictures of kitchen appliances and talk about what they do.

Links to home
- Ask parents to reinforce safety in the kitchen especially with electrical appliances.
- Ask parents to help their child to cut out pictures of kitchen appliances from a catalogue to bring to the setting for Show and Tell.

Topic
Homes

Resources
- *Doing the Washing* by Sarah Garland (Puffin)
- Catalogues with kitchen equipment such as washing machines, tumble driers, dishwashers, kettles
- Labels for equipment
- Blu-tac®

⚠ Tell the children that they must never touch electrical appliances unless an adult gives them permission.

Put out the fire

Topic
People who help us

Resources
- *Pixie* (Swallow Systems) see www.swallow.co.uk
- Boxes
- Red shiny paper
- Plastic tubing
- Syringes
- Water
- Small world figures

Learning objectives
- To develop and demonstrate skills in making a technological toy work
- To play cooperatively, taking turns and sharing fairly with others
- To work within a group, taking account of one another's ideas about how to organize the activity.

Preparation
- Familiarize yourself with the programmable toy.

What to do
Outside activity
- Lay out a course, like a street, in a large space indoors or outdoors, for the 'fire engine' to move along. Add red shiny paper to a 'house' (as fire).
- Use *Pixie* (a floor robot) as the fire engine going to the fire.

Extension/variation
- Connect a plastic tube, with the syringe of water at the end, to small world figures (firemen) to put out the fire.

⚠ Ensure there is sufficient space to accommodate the computer system, the control box, leads and models as well as for moving the model around on the floor. Keep water away from *Pixie*.

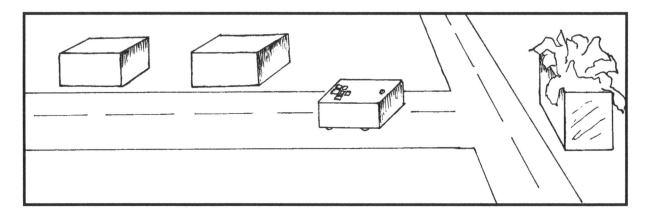

Personal, Social and Emotional Development with Understanding the World and Mathematics

A letter to Santa

Learning objectives
- To recognize that a range of technology is used in familiar equipment for everyday life in places such as homes, schools and nurseries
- To select and use technology for particular purposes
- To understand the different ways of communicating with people in the community

Preparation
- Prepare a letter to Santa on computer so children/practitioner can add a few lines underneath and write their name.

What to do
Circle time
- Read one of the *Postman Pat* stories. Discuss what a postal worker does. Ask the children what other ways they can think of to get a message to someone.

Computer activity
- Write a letter to Santa Claus using the computer. Help the children to save and print the letter.
- Show the children a fax machine if available.
- Support the learning of new vocabulary and developing language skills by introducing words such as: *letter, address, date, Dear, Love from, typing, word processor, e-mail* and *fax.*

Extension/variation
- Send an e-mail to children's television.

Links to home
- Invite parents to write down their e-mail addresses or fax numbers for their children, so that their own children can send them short messages.

Resources
- *Postman Pat* stories by John Cunliffe (Hodder Children's Books) – also available on DVD and on CD
- CD recorder
- Computer with e-mail access
- Fax machine
- Stamps
- Envelopes and writing paper

The lights have gone out!

Topic
Homes

Resources
- *A Dark Dark Tale* by Ruth Brown (Mantra Publishing)
- Christmas tree lights
- Table lamp
- Plastic torch to dismantle
- Materials to make electric circuit as shown (or ready-made units)
- Large tray
- Torch

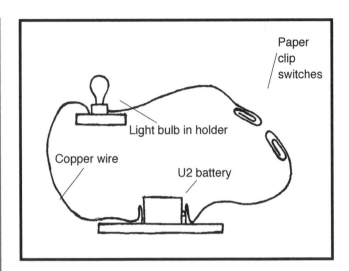

Paper clip switches
Light bulb in holder
Copper wire
U2 battery

⚠️ **Caution** with batteries as they can burn if put in the mouth.

Keep a check of bulbs – young children could break them or even put them in their mouth.

Work in a tray to reduce the chance of bulbs rolling off the table and breaking on the floor.

Learning objectives
- To find out about, and identify some features of electric circuits and investigate how they work
- To know how to operate simple equipment
- To select and use technology for particular purposes

Preparation
- Remove one bulb from the Christmas tree lights.
- A practitioner will need to sit near to an electric socket for this activity.
- Place table with electric circuit equipment on carpeted area.

What to do
Circle time
- Read the story *A Dark Dark Tale*. Ask the

children why it was so dark. Ask them what they could do to make some light.
- Show them a table lamp and ask them how to make it work and light up. They will probably suggest switching it on or inserting the plug. Show the Christmas lights and switch them on. When they do not light, replace bulb.

Table activity
- Support the children in their investigations with electric circuits.
- Talk about safety with electricity, and dangers of places where there is the danger sign (transformers).

Extension/variation
- Draw the electric circuit that lights up, and the one that does not light up.

Science notes
- Electricity flows through metals because the electrons can 'jump' from one atom to the next as they are closely packed. A battery supplies the driving force to 'push' the electrons along the wire. The thin metal wire inside the bulb becomes heated and glows. Hence electric energy is changed to heat and light energy. If there is a break in the circuit (caused by a broken wire in the bulb, or an open switch) then the electrons (electricity) cannot flow.

Related activity
- Help me to cross the road (see page 91)

All about the lion dance

Learning objectives
● To select and use technology for particular purposes, to support learning
● To manipulate a mouse, coordinate actions and find out how to access information from the Internet
● To capture experiences and responses with a range of media, such as music and dance, paint and craft materials

Preparation
● Check that information on the Chinese New Year and lion dance is accessible from the Internet.
● Find out the date of the Chinese New Year.

What to do
Computer activity
● This activity will involve a practitioner giving full support at the computer. Show the children how to access the web page, and discuss the information displayed on the different rituals.

Circle time
● Read *Lion Dancer* and *Ernie Wan's Chinese New Year.*
● Talk about the Chinese New Year to the children and organize them into appropriate groups for activities (See other activities related to the lion dance.)
● Perhaps those children who have learned about the rituals from the Internet would tell the rest of the class about them.

Extensions/variations
● Play a CD of traditional Chinese music that can be used for the lion dance performance.
● Video the lion dance performance.
● Encourage the children to work cooperatively to create a dance to the music.
● Provide resources and support them in making appropriate costumes for the lion dance. Each costume should fit as many children under it as possible.
● Practise with the children and prepare the lion dance for performance for an audience.
● Film the lion dance performance and put it onto a DVD. Watch it with the children at a later date.

Topic
Celebrations

Resources
■ *Lion Dancer* and *Ernie Wan's Chinese New Year* by Kate Waters and Madeline Slovenz-Low (Scholastic)
■ Computer with Internet access
■ Search for lion dance on Internet
■ Video camera
■ CD of traditional Chinese music

⚠ Check web pages first to determine their suitability.

Links to home
● Ask if a parent would be willing to film the lion dance performance, so that all practitioners may be involved with the children.
● Invite all parents to watch the lion dance.

Mathematics

Numbers

- Children learn number names and their sequence through rhymes, songs and stories, through counting objects during play activities, such as building bricks into a tower or sorting items into groups by colour or size, or by matching numbers to practical experiences, by counting the steps as they climb the stairs or the buttons on a coat as it is fastened. From understanding how to compare quantities simply, as 'lots' or 'a few', children begin to correctly select small numbers of objects, when asked to choose one, two or three, and to know that quantities change when something is added to a group or taken away.

- Practitioners can help children to develop their understanding of numbers by singing and acting out counting songs, playing games involving counting or addition and subtraction, such as board games or hopscotch and skittles, and introducing scoring during activities and challenges. They should support children in learning and fully understanding one-to-one correspondence by pointing to each object as it is counted and saying the number name aloud.

- Children should be encouraged to count natural materials outdoors, as well as toys and equipment indoors, and to use small toys, buttons, counters, pencils or their fingers as counting aids. They also enjoy counting steps, jumps or hops while playing the games 'What's The Time Mr Wolf?' and 'Simon Says', or finding out how many steps travel the length of the climbing frame and how many jumps carry them from one side of the room to the other.

- Once they can count confidently, children begin to recognize some numerals, such as their own age on birthday cards and the house number on their front door, and to try to copy and write them or to record them using their own marks. Adults can support and encourage them by providing number books and items marked with numbers, such as rulers, tape measures, calculators and telephones, for them to use and copy from. Children should be encouraged to count objects to match and check numerals and to trace over them in books with their fingers. Writing them in the air and in sand trays and cutting them out of playdough also helps to reinforce the shapes.

Personal, Social and Emotional Development with Understanding the World and Mathematics

© Mavis Brown and Rebecca Taylor
Brilliant Publications

Practitioners should join in with the play and model the correct formation of the numerals and how they should look when children are writing numbers.

- Numbers can be introduced into all areas of provision to check children's understanding and confidence. The aim for the end of the foundation stage is for children to be able to count reliably from 1–20, place numbers in order correctly and say which number is one more or less than the one given. With counting aids, children should be able to add or subtract single digit numbers and solve problems involving doubling, halving or sharing.

- Counting before seeking when playing 'Hide and Seek', sharing out cards or pieces before playing a game and deciding how many snacks would be needed for a group are practical ways for children to develop and use their mathematical skills.

**Personal, Social and Emotional Development
with Understanding the World and Mathematics**

Animals in sand and water

Topic
Animals

Resources
- Variety of plastic animals
- Plastic trays
- Sand
- Jug for water
- Food colouring
- Washing-up liquid
- Pebbles

⚠ If any children have allergies to baby bath or severe eczema, use only plain water.

Learning objectives
- To count reliably with numbers 1–10
- To solve problems by counting on and counting back
- To play cooperatively, taking turns and sharing fairly

What to do
- Place some animals in your sand tray, such as elephants, crocodiles, zebras and camels.
- Join in with a small group of children as they play in the sand, asking questions such as, *How many elephants can you count?*; *Can you dig a hole for four camels?*; *Can you dig a stream for a crocodile to swim in?*
- Make your water area a fun and exciting place to play by adding food colouring to the water and baby bath to give it bubbles.

- Encourage the children to count how many pebbles they use to make an underwater fun park for the plastic fish.

Extension/variation
- Set the children a theme to their play, such as *Can you make a castle for the elephant?*; *Can you make a playground for the zebras?*

Links to home
- If children have sandpits at home, encourage their parents to play counting games with them and to create numbers in the sand, using fingers, sticks and the edges of spades.
- Encourage parents to make time spent in the paddling pool and in the bath a fun learning experience, by providing an assortment of toys and containers and asking questions such as:
  - ◆ How many cups of water will I need to fill this bowl?

Colour towers

Learning objectives
- To count reliably with numbers 1–20
- To name shapes and use mathematical language in play
- To count or select up to six objects from a larger group
- To count an irregular arrangement of up to ten objects
- To recognize, create and describe patterns
- To show good control in large and small movements

What to do
- Present the children with a box of coloured blocks.
- Ask the following questions and support them in carrying out tasks:
    - Can they count out of the box six red blocks?
    - Can they carefully build a tower with the blocks?
    - Can they count out ten yellow blocks?
    - Can they make a repeating-pattern tower by selecting two colours of blocks?
- Suggest that children make an irregular arrangement of blocks on the floor and ask friends if they can estimate how many are there and then check by counting them.
- Having given a number theme to their play, encourage them to ask each other to make towers using different numbers of bricks.
- Have the children put an irregular arrangement of blocks on the carpet and get their friend to first estimate how many are there and then count them.

Extensions/variations
- In your outside area have big wooden blocks that the children can use to make structures.
- Ask questions such as *How many bricks do you think we need to make a house?*
- Make comments such as *I wonder how many wooden bricks there are altogether.*

Topic
Colours

Resources
- Coloured blocks
- Large wooden blocks to use in outside area
- Books about colours to be stored in reading corner

Links to home
- Encourage parents to take every opportunity to count with their children at home. They could count their Lego® bricks or other toys as they put them away, the teddies on the beds, the wheels on the cars, the stairs or how many steps it takes to get from one side of a room to another.

Animal hospital

Topic
Animals

Resources
- Three-sided screen
- White backing paper
- Staple gun, staples
- White and red card
- 10 animals
- 10 cushions
- 6 white adult shirts with the sleeves trimmed to make them manageable for the children
- Large pieces of card
- Thick black felt tip pen
- Bandages of different lengths made out of an old white sheet cut into strips
- Medical bag
- White boards, wipeable pens
- Telephone
- Appointment book
- Table and 2 chairs for the reception desk

will help adults interact with the children in the hospital, such as: *How many patients are there in the hospital today?; How many patients have poorly heads?; How many patients have poorly arms or legs?*

- Work alongside the children, helping them to select the right length bandage for the right job. An arm for example may need a shorter bandage than a head.
- At the end of the session encourage the children to tidy up by checking that all the white shirts are hanging on their hooks, that each 'patient' has its own cushion and that they are comfortable for the night ahead.

Learning objectives
- To count reliably with numbers 1–10
- To count and check in a role-play situation
- To compare and order items by length
- To record numbers, using recognizable numerals and other marks that can be interpreted and explained
- To play alongside other children who are engaged in the same theme

What to do
- In your role-play area, invite the children to help you set up an animal hospital. Collect the resources listed and staple the white backing paper on your screen to create a hospital environment. Make nurse's hats for your children out of strips of white card; cut out a red cross and glue it on the front.
- Write up questions on big pieces of card that

Extensions/variations
- Some children will be able just to tell you how many animals there are in the hospital whilst others might be ready to record the number.
- Provide them with white boards and wipeable pens, so that they can write any numbers and other marks that they choose and rub them out and change them as they play.
- Show some children how to tally as they count.

Links to home
- Find out if any of your parents are nurses who might be prepared to talk to the children about their job and even show how people can be bandaged.

Birthday shop

Learning objectives

- To use numbers confidently within a role-play situation
- To relate numbers to ages and quantities, recognize numerals on cards and record numbers of items required
- To use everyday language to talk about money and to handle coins and solve problems
- To recognize and describe special times and events for family and friends

What to do

- Gather the children in front of you and ask them to tell you what they think a birthday shop should sell. Write the items as a list on your easel. Count with the children to find out how many items they have thought of.
- Gather a group of willing helpers and set up your birthday shop in your role-play area.
- Set up your three-sided screen and staple white backing paper on to the walls.
- Encourage the children to paint balloons on the paper. Ask whether they can paint, for example, three red balloons and two blue balloons.
- Set up one table at the front of the shop and place on it the birthday cards, wrapping paper, till and money.
- At the back of the shop set up a table where the children can make birthday items such as cards, birthday hats and birthday banners.
- Put up a big sign that says 'Birthday Shop'.

Extension/variation

- Put some questions up in your birthday shop that will help members of staff interact with the children while they are working, for example: *How much are your birthday cards?*; *How many balloons do you have for sale?*; *Would you be able to make a birthday hat that will fit my head?*

Links to home

- Encourage parents to allow their children to handle and play with money at home, counting coins and making towers with them.

Topic
Celebrations

Resources
- Easel, or other board for teacher to write on
- Three sided screen
- White backing paper
- Staples
- Stapler
- Poster paints
- Two tables
- Used birthday cards
- Used wrapping paper
- Till
- Plastic money
- Balloons
- A4 card
- Envelopes
- Large card
- Large strips of card
- Felt tipped pens

- Encourage the children to talk about how they celebrate birthdays in their family and which shops they visit to buy presents and cards.

Personal, Social and Emotional Development with Understanding the World and Mathematics

Countdown to Christmas

Topic
Celebrations

Resources
- Template (page 191)
- A variety of Advent calendars
- Poster paint
- Paintbrushes
- Sugar paper
- Circles from A4 white paper
- Red card
- Templates for numbers
- Used Christmas cards
- White paper
- Templates for letters
- Pale blue backing paper
- Till receipt roll
- Glue
- Blue-tac®
- Scissors
- Staples
- Staple gun
- Drawing pins

⚠ Supervise use of scissors.

Learning objectives
- To learn that Advent is a special time of the year when we count down to Christmas
- To begin to use mathematical names for 2D shapes and to understand halves and quarters
- To recognize, create and describe repeating patterns

What to do
- Show the children a variety of Advent calendars and tell them that they are all going to work together to create a class Advent calendar.
- Work out how many days in December that you have left in your setting. You may just want to count down to your last day rather than to Christmas Eve.
- Photocopy the snowmen on page 191 for the children to colour in, or ask them to paint their own snowmen. Encourage them to paint a repeating colour pattern on each scarf, such as: red, blue, red, blue or yellow, red, green, yellow, red, green.
- Give the children some white circles and show them how to fold them in half and then into quarters.
- Help them to cut shapes out of their circle and then open them to reveal their very own special snowflake.
- Using your number templates and red card cut out the numbers that you require.
- Once the snowmen are dry cut them out and stick a number on to each one.
- Stick a picture cut from the front of a Christmas card on to the back of each snowman.
- Using your letter stencils cut out on white paper the words 'Countdown to Christmas'.
- Put up your pale blue backing paper and staple a till receipt roll all the way around the edge. Paint a zigzag all the way round.
- Blu-tac® the words on to the top of the display. Pin snowmen on to the display, mixing up the order of the numbers.

Extension/variation
- Each day in December, ask the children which number snowman you should turn around. Invite them to guess what might be behind the snowmen. Choose pictures like a Christmas tree, Father Christmas and a Christmas pudding.

Links to home
- Encourage the children to talk about what is behind the doors of their Advent calendars at home.
- Suggest to the children that they try to count how many Christmas cards they have in their house so far, how many Christmas trees, how many presents under the tree and how many baubles hanging on it.
- Explain to parents that you will be talking about Christmas in the setting and respecting that any children who celebrate a different festival, such as Hanukkah or Diwali, will be learning about the customs of the country they are living in. Some parents (such as Jehovah's witnesses) may choose for their children not to take part in certain activities.

Five little Diwali lamps

Learning objectives
- To count out 10 objects from a larger group
- To learn about the Hindu festival of light known as Diwali, celebrated in October/November
- To construct with a purpose in mind, using a variety of resources and malleable materials

What to do
- Show the children how to make a little lamp holder by moulding Plasticine into a sphere and making a hole with their thumb big enough to take a nightlight candle.
- Cut up the string of beads and show the children how to press the beads into the Plasticine. Tell them that they can only use ten beads for each lamp.
- Show them how to lightly press the base of their candle holders so that they sit on a table. Repeat until they have five little lamp holders and can count them. They will make a lovely display on their lounge windowsill.

Extensions/variations
- Many Hindus use chalks to draw rangoli patterns outside their homes on the pavements to welcome visitors. Rangoli patterns are usually very bright and symmetrical and have lots of shapes in them.
- Encourage your children to experiment in your outside area, using playground chalks to make their own rangoli patterns. They will certainly welcome any visitors to your setting.

Links to home
- Ask the children to find out if they have any candles in their homes and when anybody lights the candles. Perhaps they are lit at special family celebration dinners.
- Ask any of the children's parents who celebrate the festival of Diwali to come into the setting and talk to all the children about their customs and traditions. If none of the families celebrate Diwali themselves, ask if anybody has a friend or neighbour or another relative who does.

Topic
Celebrations

Resources
- Plasticine
- String of beads that are usually draped around Christmas trees
- Scissors
- Nightlights
- Thick playground chalks

Personal, Social and Emotional Development with Understanding the World and Mathematics

Number igloo

Topic
Seasons

Resources
- Three-sided screen
- White backing paper
- Stapler
- White sheet
- White cushions
- Thick black felt tipped pens
- A variety of soft toy penguins and polar bears of different sizes
- A4 card
- Hats
- Gloves
- Scarves

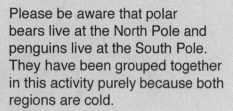

⚠ Please be aware that polar bears live at the North Pole and penguins live at the South Pole. They have been grouped together in this activity purely because both regions are cold.

- Tell the children that today they are going to create a cold, icy igloo for the polar bears and penguins.
- Set up your three-sided screen and cover it with white backing paper. Put white sheets with cushions on the floor, and a box of warm clothing.
- Tell the children that the penguins and polar bears only like eating numbers.
- Invite the children to write numbers on pieces of card, using the felt-tipped pens, and put them in the igloo.
- Encourage the children to dress up for cold weather using the clothes that you have placed in a box in the igloo. Ask the children whether each of the animals has a friend.

Learning objectives
- To use number names and language accurately in play
- To record and represent numbers using numerals and other marks
- To use objects to add two single digit numbers and count on to find the answer
- To talk about the features of animals and their environments and how different environments and seasons vary from each other

What to do
- Ask the children to think about which animals like to live in cold, icy environments. Help them to understand that penguins and polar bears are animals who live in cold areas. Encourage the children to think about what season is cold in our country.

Extensions/variations
- Make some simple addition cards, for example 2+1= , 3+1= , 5+1= , and place them in the igloo so that the children can copy them on to the walls and calculate them using the penguins and polar bears.
- Ask the children to put the animals into size order and say which is the biggest penguin and the smallest polar bear.

Links to home
- Ask parents to lend penguins, polar bears and sets of hats and gloves to put inside the igloo for the children to use. Make sure the hats and gloves are clearly labelled.

Personal, Social and Emotional Development with Understanding the World and Mathematics

Number values

Learning objectives
- To talk about and record numerals of personal significance
- To use number names accurately
- To represent numbers by forming numerals correctly in the air and then writing them down
- To handle pens effectively for writing

What to do
- Gather the children into a circle and tell them that today they are going to talk about numbers that are important to them.
- Talk about numbers which are important to you, for example your age and date of birth, your shoe size, your house number, your phone number, your car registration number, the number of people in your family, how many brothers and sisters you have, your lottery numbers or your lucky number.
- You might not want to disclose all these numbers to the children but it is important for them to know that numbers are a part of everyday life.

Extensions/variations
- Encourage some children to record by showing fingers or drawing numbers in the air.
- Ask other children to record their numbers on white boards, modelling how to form each numeral correctly by writing it first on your own white board.
- Make signs that show significant numbers in your setting, such as: *We have 6 aprons*; *We have 30 pencils*.

Links to home
- Encourage parents to teach their children their addresses and telephone numbers as soon as possible.
- Remind parents of the importance of keeping the setting informed of any changes in contact details.

Topic
Families

Resources
- White boards
- Wipeable pens
- Large easel
- A4 card
- Large felt-tipped pen

Bear washing line

Topic
Animals

Resources
- Bear washing line template on page 192
- Thick felt tipped pen
- A4 card in two different colours
- Laminator
- String
- Pegs
- Pulley mechanism (optional) for repositioning the line

Learning objectives

- To count reliably with numbers 1–10 and place them in order
- To recognize numbers 1–10
- To identify the missing number

Preparation
- Photocopy the bear template on page 192 on to card and cut it out.
- Draw around the template on alternate pieces of coloured card ten times and cut them out.
- Write a number on each tummy.
- Ensure that all the odd numbers are written on one colour card and the even on the other. Laminate the bears for durability.
- Set your washing line up across a corner of the room where it can be lowered to your children's height when in use.

What to do
- Show the children the bears, asking them what they have on their tummies.
- Pick one bear out and ask a child to peg it up on the washing line where they think it should go.

- Encourage them to use their knowledge of numbers so number 5 would be pegged roughly half way across the line.
- Once all the bears are pegged up ask the children if they are all in the right order or if any bears need to be moved?
- As a group, say the number names in order.
- Finish by taking all the bears down and hiding one. Peg the bears up placing them in order and then ask the children to guess which is the missing number.
- Repeat several times.

Extensions/variations
- Set the washing line up in your maths area alongside the units where you store your maths equipment.
- On top of your units lay out activities the children can choose to use independently during free play sessions.

Links to home
- Encourage parents to put up number friezes in their children's bedrooms or playrooms, so that the children are stimulated by seeing numbers around them both at home and in the setting.

Personal, Social and Emotional Development with Understanding the World and Mathematics

Ten brown bears

Learning objectives
- To have a secure knowledge of the order of numbers 1–10
- To join in with a number song involving counting back
- To say which number is one less than the given number
- To use available resources to create props to support play

Preparation
- Paint the box to represent a wall. Make it sturdy. If using template bears, cut them out, letting the children colour them; cut five slits in both of the long sides of the box.

What to do
- Place the cardboard box on the table in front of the children, with the bears on top or in the slots.
- Sing the following song to the tune of 'Ten green bottles':
  Ten brown bears sitting on the wall
  Ten brown bears sitting on the wall
  And if one brown bear should accidentally fall
  There'll be nine brown bears sitting on a wall.
- After each bear has fallen, stop and ask the children to work out how many bears there are left by counting them.
- As the children sing, encourage them to show the correct number of fingers to represent the bears left on the wall.

Extensions/variations
- You may decide to make ten bear hats that the children can wear. Help the children to paint strips of card brown. Staple the ends together to fit the children's heads and attach brown ears.
- When the bear hats are not in use peg them on to a washing line in your play area.

Topic
Animals

Resources
- 10 brown bears or 10 cardboard bears from the template on page 193
- Cardboard box
- Paint and brushes
- Felt-tipped pens or crayons

Links to home
- Ask the children to sing the song at home with their parents, acting it out with their toys. Encourage parents to sing lots of counting songs with their children – it's something they can do wherever they are, such as in the car, shopping or just walking.

Pass the animal

Topic
Animals

Resources
- A soft animal of your choice
- Ball

Learning objectives
- To recite numbers 1–10 in order, starting from any number
- To count on and count back
- To cooperate with each other, taking turns and sharing fairly
- To follow instructions involving several ideas or actions

What to do
- Ask the children to sit in a circle.
- Explain to the children that they are going to pass the toy around the circle, counting each pass together.
- Allow the children to pass the animal around while you count up to 10. You can then start again, encouraging them to join in.
- As the children become more confident, start the game by passing the animal but use a different start number other than the number 1, for example 3 or 4.
- Encourage the children to get quicker as they pass the animal.
- You could always clap your hands to encourage a quick response and steady beat.

Extensions/variations
- For higher ability children you may start on number 5 and encourage them to take away one number instead of adding, for example 5, 4, 3, etc.
- Make the game more energetic by taking it outside and, instead of passing an animal, use a ball. As the children become more confident at passing the ball they could throw it to each other as they count.
- Try to have a number session at the same time every day so that it becomes part of the children's routine.

Links to home
- Encourage the children to repeat the activity at home, with all their family members joining in the counting.
- Invite them to think about where they and their friends or family could practise counting, for example around the dinner table, or in the car or lounge.

Personal, Social and Emotional Development with Understanding the World and Mathematics

Let's get physical

Learning objectives

- To count actions or objects that cannot be moved
- To measure short periods of time in simple ways and with non-standard units
- To move freely and with pleasure and confidence in a range of ways
- To be able to count things they cannot touch

What to do

- This activity can be used on its own or as a quick five-minute warm up with the children before you start any physical activity.
- Encourage them to keep moving with the following ideas:
  - ◆ Ask them to do five jumps on the spot.
  - ◆ Can they do three star jumps on the spot?
  - ◆ Can they do ten marches on the spot?
  - ◆ Can they bend down and reach for their toes four times?
- Turn the sand timer over and ask the children to count how many times they can jump up and down before all the sand runs through.
- Encourage the children to share your large toys outside fairly, using time as a form of measurement. For example, you can ride around the playground on the big bike three times and then you must give the bike to anyone who is waiting for it.

Extensions/variations

- Children can gain a better understanding of measurements and amounts by counting while taking part in practical activities, so talk to them about what you are doing and encourage them to count along with their actions. For example: *We are going to do three steps forward and then three steps back* or *We are going to lift our knees up five times.*
- Make it even more exciting by doing it to energetic pop music.

Topic
Health

Resources
- ■ Sand timers
- ■ Large bikes
- ■ Energetic pop music

Links to home

- Ask the children to count how many stairs they have in their houses and whether they have any steps in their gardens or up to their front doors.

Personal, Social and Emotional Development with Understanding the World and Mathematics

Body bingo

Topic
Myself

Resources
- Templates on pages 194 and 195
- A3 paper
- Felt-tipped pens
- Clear laminate sheets
- Scissors
- Plastic folder with zip
- Glue
- Beads

Learning objectives
- To recognize numbers 9–14
- To count dots on a die
- To understand that arranging objects helps counting and recognition of numbers and amounts
- To play cooperatively, taking turns and sharing fairly
- To work as part of a group, understanding and following the rules

Preparation
- Photocopy page 194 onto A3 paper six times. Colour with felt-tipped pens if you want to, then laminate each sheet and cut out the corresponding shapes that you will fit on to the game boards. Photocopy on to thin card the die template on page 195 and laminate it. Store all the bits for the game in a plastic folder with a zip.

What to do
- Invite six children to sit around a table or on the floor to play a game. Give them each a game board and lay out all the corresponding shapes around them.

- Encourage the children to take turns to throw the die, count the dots and place the right card on to the board.
- Ask the children if they can name the shapes as they pick them up.
- The first child to collect all their shapes wins the game by shouting 'Body bingo!'

Extension/variation
- Give the children ten beads and show them how to put them into two lines of five to help them count more easily. Give them eight beads or four beads and ask them to try this for themselves.

Links to home
- You should by now be building a good store of maths games. Your children may borrow books to take home but you might now want to consider whether they can also borrow a maths game to play with their parents. Write a short instruction card to accompany each game.

Gingerbread men

Learning objectives

- To count an irregular arrangement of up to ten objects
- To solve practical problems involving halving and sharing
- To find the total number of particular items by counting all of them
- To measure out and compare quantities in a practical situation
- To handle tools and malleable materials safely and with increasing control
- To manage own basic hygiene and appreciate the need for safe practices with regard to hygiene for good health

What to do

- Remind the children to help and support each other in fastening aprons and washing hands before beginning the cookery activity.
- Spoon the honey and syrup into the saucepan and heat gently with the margarine.
- Put the flour, ground ginger and egg yolk into the bowl and stir in the honey mixture.
- Mix the bicarbonate of soda with the cold water and add to the mixture.
- Encourage the children to take turns to knead the dough on a lightly floured surface until smooth. Roll the dough out thinly and encourage each child to use the pastry cutter to cut out a gingerbread man. Place the gingerbread men on two lightly greased baking trays and cook in the oven for 10 minutes at gas mark 4 or 180°C.
- Allow the gingerbread men to cool and then ask the children to share them out.
- Discuss how to do this, considering the following questions: *Are there enough for everyone? If not, how are you going to share them? Which way should you cut them in half?*

Extension/variation

- Before the children cut and share out the gingerbread men, ask them to count how many arms and legs they have in total.

Links to home

- Suggest to parents that they might buy or borrow the story of the gingerbread man, available in *Storytelling with Puppets, Props*

Topic
People who help us

Resources
- 2 tblsp clear honey
- 1 tblsp golden syrup
- 25g of margarine
- 175g plain wholemeal flour
- 1 teaspoon bicarbonate of soda
- 1 teaspoon cold water
- 2 teaspoons ground ginger
- 1 egg yolk
- Scales
- Tablespoon, teaspoon
- Wooden spoon
- Bowl
- 2 baking sheets
- Saucepan
- Rolling pin
- Gingerbread man pastry cutter
- Large serving plate
- Aprons

 Check for allergies. Let the gingerbread men cool before allowing children to share.

and Playful Tales (Brilliant Publications) or from their local library and count with their children how many characters were chasing the gingerbread man.

Personal, Social and Emotional Development with Understanding the World and Mathematics

Bunny boxes

Topic
Celebrations

Resources
- Large sheets of white card
- Bunny box template on page 196 (draw around the templates and cut them out for the children, as they are quite fiddly)
- Crayons
- Pencils
- Scissors
- Circles of tissue paper
- A large bag of mini eggs to share amongst the children

- Put the boxes on a special Easter table and tell the children that if they are very good the Easter bunny might come along and fill them. You will find that the children will keep checking their boxes.

Learning objectives
- To count reliably with numbers up to 5
- To experience how shapes slot together
- To use developing mathematical ideas and methods to solve practical problems
- To enjoy joining in with special seasonal events
- To manipulate materials to achieve a planned effect

What to do
- Make a bunny box in advance, using the template on page 196, to show the children what their boxes will look like.
- Help them to decorate the bunnies, reminding them that each one needs two eyes, one nose and one mouth.
- Ask them to turn the boxes over and repeat on the other sides, then decide how many bunnies there are altogether in each box.
- Show them how to fold the bunnies along the dotted lines and then how the arms slot together.
- Encourage the children to choose their favourite colour tissue paper and put it inside the box.

Extension/variation
- On the last day before Easter in your setting put the big bag of mini eggs on your Easter table with a note from the Easter bunny. Tell the children that you do not know how to share them fairly and invite them to make suggestions. Any variation on putting one into each basket and then another into each basket until there are not many left could be acted upon. A practitioner should demonstrate the sharing out while the children watch carefully, to ensure that they all understand and believe that it is fair.

Links to home
- If there are lots of small eggs, encourage the children to share the eggs with their families and to work out how many each family member could have.
- Encourage parents to arrange an Easter egg hunt for their children at home, or even a sock and shoe hunt. Explain to parents that it is a great way to develop the children's counting skills.

Personal, Social and Emotional Development with Understanding the World and Mathematics

Colourful cube game

Learning objectives
- To count reliably with numbers 1–6 and possibly further
- To recognize and talk about numbers 1–20 and possibly beyond
- To use one-to-one correspondence to move a counter correctly
- To recognize and respond to the numbers on a die by moving the appropriate number of places
- To know that a die is a cube
- To play cooperatively, taking turns and not needing to 'win' the game

Preparation
- Photocopy the game board on page 197 twice. Colour one copy, making the squares a mix of red, yellow and blue. Write numbers 1–29 in order onto the other copy, beginning with the square after 'start'.
- Laminate them.
- Photocopy the die on page 198 twice. Colour the appropriate circles in the three colours on one copy. Write numbers 1–6 in the circles on the other copy. Laminate them and fix them together with sticky tape.

What to do
- Gather a small group of children together, explaining that today they are going to play some board games.
- Show them the coloured die and the coloured board.
- Children take turns to throw the die, say which colour it has landed on and move the counter to the next square of that colour. The aim is to get to the end of the track first.
- Next, show the children the numbered die and board and ask them to play the game again, this time saying which number the die has landed on and counting how many squares to move along the track. They should notice that the numbers on the board get larger, but focus on counting one square at a time to move up to six squares along the track with each turn.

Extensions/variations
- If you are able to paint your outside area, large board-game designs such as snakes

© Mavis Brown and Rebecca Taylor
and Brilliant Publications

Topic
Colours

Resources
- Template of board game on page 197
- Template of cube on page 198
- Felt tipped pens
- Laminator
- Store all your games in your maths area and ensure an adult always joins a group when a game is being played to encourage fair play and positive behaviour

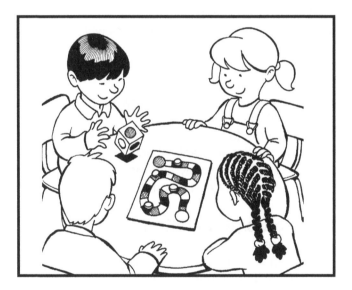

and ladders always interest the children, but make sure you use special outdoor paints.
- Provide the children with a foam die and encourage them to be the counters.
- Paint a hopscotch board and perhaps a number caterpillar that the children can move along.
- If painting is not possible, you can always make grids for the children in chalk.

Links to home
- Encourage parents to play board games with their children at home, for example snakes and ladders, Ludo and Mousetrap®.

**Personal, Social and Emotional Development
with Understanding the World and Mathematics**

We have a pet tiger

Topic
Animals

Resources
- A stuffed toy tiger
- A basket big enough for the tiger to lie in
- Cubes or buttons, or something that the children find easy to count
- A sign saying 'Shh the tiger is sleeping' (place this in the basket)

Learning objectives
- To estimate how many objects while looking at them and then check by counting them
- To recognize a mistake in counting and be able to correct it
- To have a secure knowledge of the order of numbers
- To begin to understand addition and subtraction in practical activities
- To listen attentively and respond to what is heard with relevant comments and answers

Setting up/preparation
- Place the tiger in his basket in your maths area so that the children begin to associate him with numbers and fun.

What to do
- Gather the children in front of you and sit down with the tiger on your lap.
- Tell them that today you are going to help the tiger to learn his numbers.
- Hold in your hand five buttons.

- Encourage the children to predict how many buttons are in your hand.
- Count them together to see if anyone's prediction was close.
- Ask the children to cover their eyes. Tell them that the tiger is going to take some of the buttons and that they are going to try to work out how many buttons he has taken.
- At first pretend the tiger has taken just one button, so you show the children the four buttons left in your hand. Work up to him taking two, three or more.

Extensions/variations
- Suggest that the tiger could take turns with the children when they are counting. For example, the tiger would say 'two' and they would say 'three' the tiger would say 'four' and they would say 'five'. Make sure the tiger makes mistakes sometimes so that the children have to correct him.
- Ask the tiger questions such as, *Which number comes before three?* Make sure he makes many mistakes.

Links to home
- Explain to parents about the new pet toy tiger and suggest that they and their children might like to try to think of a name for him.

Personal, Social and Emotional Development with Understanding the World and Mathematics

Our tiger keeps eating numbers!

Learning objectives
- To count reliably with numbers 1–10, place them in order and know when one is missing
- To compare two numbers and use language such as *more than* and *less than*
- To find one more or one less than a number from 1 to 10
- To be confident to try a new activity, to speak within a familiar group and to talk about own ideas

What to do
- Introduce your tiger to the children.
- Say that today your tiger is very hungry and he has just whispered to you that he really likes eating numbers.
- Show the children your large number cards. Ask them to tell you the names of the numbers.
- Say that you are going to give one of the cards to the tiger and then they are going to look at the other cards that are left and decide which one is missing.
- Give a card face down to the tiger so that the children cannot see which one it is. Then lay out all the cards face up and see if the children can identify which one is missing.
- Encourage the children to explain how they worked out which number the tiger had eaten.

Extensions/variations
- Show the children a number and explain that you are going to ask the tiger to say what the following number is. Make the tiger say the wrong number so that the children have to shout out to correct him.
- Repeat several times.
- Put two numbers up on your easel and get the tiger to point to the number which is greater or smaller, saying: *Which number is more than, or less than, this one?*
- Encourage the children to think about whether the tiger is right.

Topic
Toys

Resources
- Your toy tiger (preferably with mouth that opens)
- Large number cards
- White cards numbered 1–10

Links to home
- Provide the children with a set of white cards numbered 1–10 to take home. Write down some activities that the children can do with their parents, such as putting them in the right order or hiding one and trying to guess which one is missing.

Combining groups of flowers

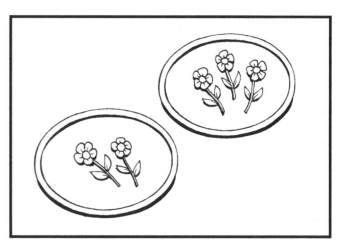

Topic
Gardening

Resources
- 2 hoops
- A variety of artificial flowers
- A soft cat toy

Learning objectives
- To count an irregular arrangement of up to ten objects
- To find the total number of items in two groups by counting all of them
- To begin to use addition and subtraction with support, in practical activities and discussion

What to do
- Gather the children into a circle and place two hoops in the middle.
- Put three flowers in one hoop and two flowers in the other. Encourage the children to work out how many flowers there are altogether.
- Repeat several times with different numbers of flowers.
- If a child feels confident enough, let them put different numbers of flowers into each hoop and then ask the other children to give the answer.
- Now just put one hoop into the middle of the circle. Introduce your naughty pussy cat, who is going to take some of the flowers. Show the children how many he has taken from the hoop and then ask them how many are left. Repeat again with the pussy cat taking different numbers of flowers.

Extensions/variations
- Draw on your easel two cats. Say to the children that you want them to imagine that two cats went into their garden and ask them to think about how many cats' legs they would be able to see in total, and how many cats' tails.
- Draw on your easel two people and a dog. Say to the children that they were looking out into their garden and saw two friends and one dog and ask them to think of how many legs they would have been able to see in total.

Links to home
- Encourage the children to try out number problems on their parents at home. For example, *If I had three chocolate bars and you had two, how many would we have altogether?*

My family number game

Learning objectives
- To count objects that cannot be moved, such as spots on a die
- To begin to develop an instant recognition of one, two or three spots on the die
- To play cooperatively, taking turns and supporting each other

Preparation
- Photocopy the number track and die on pages 199–200 for every child.
- After the child has coloured in the dots on the die, laminate it to make it last longer, cut it out and glue it together.

What to do
- Provide each child with a number track and encourage them to decorate them with members of their family. Perhaps at the finish they could draw their house.
- Some children may be able to trace over the numbers with a felt-tipped pen whilst others might need you to do it for them, but encourage them to watch while you do it.
- Play the game in small groups, helping the children to work out what number the die lands on, then to move that number on the board.

Extensions/variations
- Play dominoes with your children. At first they might have to count the spots but as they gain more experience, they will be able to instantly recognize one, two or three.
- Encourage the children to look at the pattern that four dots make and see the square shape, to recognize that the five dots look a bit like a face and the six dots like two legs.

Links to home
- Provide the children with a carrier bag and let them take their family number game board home. If possible let them take their die as well in case they do not have one at home.
- Encourage them to play the game with their parents and their older brothers and sisters as well.

Topic
Families

Resources
- Board game template on page 199
- Die template on page 200
- Glue
- Laminator
- A box of dominoes
- Felt-tipped pens

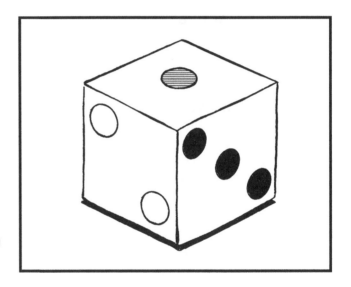

My number book

Topic
Myself

Resources
- Little books
- Pencils
- Crayons
- Long strips of sugar paper
- Stapler and staples

- On the third page, you might get them to draw around their foot and then write their shoe size inside it.
- Some children might like to finish by writing page numbers.

Extension/variation
- Make hats for the children from long strips of paper. Encourage each child to write a number on her hat that is significant to her, such as her age or her house number. Measure the length of paper you need for each child's head, then staple her hat to fit and help her to cut triangles out of the top to make a crown. Arrange a hat parade and ask all the children to guess why their friends have particular numbers on their hats.

Links to home
- Encourage the children to take their books home to share with their parents and ask whether they can think of any other significant numbers to add to the pages.

Learning objectives
- To select the correct numeral to represent numbers
- To recognize numerals of personal significance
- To use numbers as labels
- To record numbers, using numerals and other marks
- To handle equipment and tools effectively, including pencils for drawing and writing

What to do
- Provide the children with a little blank book and tell them that they are going to make a number book all about themselves.
- On the first page, suggest that they draw themselves holding a balloon. In the balloon, ask them to write their age. Some children might be able to do this independently whilst others might need to trace over your writing.
- On the next page, suggest that they draw a picture of their house and write their house number in the middle. You might need to refer to records to check their house number.

Mathematics

Shape, Space and Measures

- Children learn to recognize first 'big' and 'little' and then degrees of size and weight through handling objects and describing pictures in books. They begin to categorize and order items by length, height, weight or capacity, comparing and talking about size in everyday language.

- They explore shapes when they play with puzzles and blocks and notice and describe patterns. They make arrangements and models with objects, take an interest in shapes in the environment, notice similarities and create their own patterns. They are interested in learning mathematical names for both 2D and 3D shapes and appropriate language to describe their characteristics.

- Filling and emptying containers in sand and water play encourages an understanding of capacity. This can be further developed through activities such as cookery, painting and other crafts.

- An understanding of position and distance begins when children create roadways and railway tracks, climb and balance across equipment, dance or run around in groups. They should be encouraged to use positional language when moving play pieces or themselves to develop a full understanding of what it means to be 'in front' or 'behind', 'on top', 'underneath' or 'beside'.

- Familiar routines and sequences of actions help children to think about time. First, they learn to anticipate events, such as meals, stories or going home, and then to understand words like 'later', 'soon' or 'before that'. With support, children should develop the ability to measure short periods of time for themselves. For example, they might count or use an egg timer to show when their turn comes. They will also learn to talk about the past and the future, explaining familiar events in a logical order and using the correct forms of speech to make their meaning clear.

- Experiences such as outings to the shops and shopping role-play introduce the concept of money and allow children to begin to learn about the names and values of coins and notes and how much some items might cost.

- Practitioners should seek to provide materials and resources, activities and experiences that encourage children to explore and talk about shapes and the spaces between them, to compare, order and measure objects in a range of ways and to develop confidence in using mathematical language.

Family journeys

Topic
Families

Resources
- Photographs
- Letter to parents
- Travelling toy
- Display board
- Display table (see page 201 for display ideas)
- Backing paper
- Till roll for border
- Poster paints
- Scrapbook
- Maps
- Globe
- Postcards
- Pictures

good understanding of the duration of time might say, 'We left after breakfast and we did not get there until bedtime. It was a very long journey.'

- You might like to create a graph with the help of the children, which shows how many times the toy has been in a plane, car or train, or has walked.

Learning objectives
- To use everyday language to talk about distance and time
- To measure periods of time in simple ways
- To count reliably with numbers 1–20 and record information in a simple graph
- To talk about the features of environments and how environments vary from each other
- To know about similarities and differences in relation to places and objects

What to do
- Introduce the travelling toy to your children by showing them photographs of you with the toy at local places of interest. It might be the local park or forest, or it might be at your house.
- Describe each journey explaining the mode of transport and how long it took.
- Send the letter home to parents, explaining that the children can borrow the toy.
- When the children come back with the toy, encourage them to talk about their journey and how long it took them. A child with a

Extension/variation
- Set up a display board in your setting which is dedicated entirely to the travelling toy. Place on it maps, postcards and photographs that describe the journeys that it has made. Put a display table next to it, where you can place a globe, and perhaps artefacts that the toy might bring back from its journey, such as shells, ornaments and guide books. Write in a scrapbook every time a child borrows the toy.

Links to home
- Letter home to parents.

  Dear parents,
  We have a very special toy that likes to go on family journeys. Your child is very welcome to borrow him and take him on holiday or to visit relatives. If it is possible, we would be very grateful if you would take a photograph of your child and the toy at your destination.

Personal, Social and Emotional Development with Understanding the World and Mathematics

© Mavis Brown and Rebecca Taylor
Brilliant Publications

Family photographs

Learning objectives
- To use positional language and describe relative positions
- To understand the vocabulary of size
- To use a range of tenses when describing sizes and positions

What to do
- Send a letter home to the children's, parents, asking for their child to bring in a family photograph on a certain day.
- Gather the children into a circle and ask them to each hold up their photograph.
- Ask children to walk around the circle showing their photograph.
- Then ask them to describe it to the other children, using vocabulary such as: *I am next to my Mummy, My Daddy is behind me, My dog is in front of me and the umbrella is above us all.*
- Ask questions to encourage them to describe the size of people in their photograph: *Who is the biggest?*; *How many people are in the photograph?*; *How many ladies and how many men?*;and *How many children?*
- Display the photographs on a board with a sign saying 'Our families'.

Extensions/variations
- Ask the children to try to draw their family photograph, making sure they include all their family members.
- Remind the children to think about size while drawing their pictures and to try, for example, to make their daddies look bigger than themselves and their dogs look smaller.

Links to home
- Be very sensitive of family circumstances. Everybody's family is special. Some families consist of just two people, whilst others may have three, four or more people in them.

Topic
Families

Resources
- Letter home to parents
- Children's photographs
- White paper
- Pencils
- Crayons
- Display board
- Card
- Thick felt-tipped pen

Shape collages

Topic
Shapes

Resources
- Sugar paper
- Templates of shapes
- Scissors
- String
- A variety of sticky circles, squares and triangles all different sizes
- Glue
- Hole punch

Learning objectives
- To show interest in shape and space by playing with shapes and making arrangements
- To use shapes appropriately for tasks
- To explore and compare characteristics and sizes of shapes and use mathematical language to describe them
- To use mathematical names for 2D shapes
- To select appropriate resources and adapt work where necessary

Preparation
- Cut uniform circles, triangles and squares from pieces of sugar paper.

What to do
- Ask the children to find something in your room that is the shape of a circle, triangle and square. Tell the children that today they are going to make shape collages.
- Show them the sticky shapes that they are going to use and talk about how some are big and some are little.

- Provide them with a piece of sugar paper that has been cut into a circle and encourage them to select only circles to stick on to it. Encourage them to stick shapes on to both sides, to make a mobile. Repeat with the triangle and the square. Hole-punch the top of the mobiles, put string through the holes and hang them on your washing line.

Extension/variation
- Put all your maths jigsaws on a shelf in your maths area that is accessible to your children. Have a particular time in the day when you do jigsaws to help your children recognize shapes and have experience of orientating them.

Links to home
- Encourage the children to make a shape collage at home using a wire coat hanger and string. They could cut out of coloured paper squares, triangles, circles, oblongs, hexagons, stars, hearts and diamonds. Their parents could help them cut the string and tie the shapes on to the hanger.

Setting up a toy shop

Learning objectives

- To use numbers and talk about money in role-play situations
- To compare quantities and objects and solve problems
- To decide on prices for items
- To handle coins, counting out payments and change
- To write numerals for price labels
- To use the language of shopping and learn new vocabulary
- To attempt to read and write appropriate shop signs

Topic
Toys

Resources
- A variety of toys
- 2 tables
- Till
- Chairs
- Paper carrier bags
- Telephone
- A4 card
- Sticky white labels
- Large felt tip pens
- Camera

What to do

- You may have a selection of toys or you may like to ask the children to bring in some toys to help you set up your toy shop.
- Invite the children to help you to lay out the toys on the table. Encourage the children to sort them and to decide which toys should be together or separate.
- Help them set up the till and the carrier bags and the telephone.
- On A4 card write some signs to go in the shop such as *SALE; Would you like a carrier bag? Please ask if you can't find the toy that you require.*
- Encourage the children to use the white sticky labels and the thick felt-tipped pens and price the toys themselves. Ask them questions, for example, *How much is the teddy?, Is the ball more expensive than the car?*

⚠ Use paper carrier bags to avoid the danger of suffocation.

Extensions/variations

- Take photographs of the children working in the toy shop. Display them – parents will find them interesting and the children will love to look at them.
- After the children have worked in the shop always encourage them to talk to the other children about what they have done and what role they took on, such as shop-keeper or customer.

Links to home

- You may like your children to find out what toys their parents used to play with when they were little. If the parents still have the toys, find out how old they are now and compare this with how old the children's toys are.

Hide and seek

Topic
Toys

Resources
- A variety of toys
- Cardboard boxes
- Large sheets of
  material to make dens
- Stop watch or egg timer
- Plastic numbers
- Wooden numbers

- Encourage them to hide a toy and then give instructions to their friends on how to find it without giving away its location immediately.
- Talk to the children about how you can also say 'warm' and 'hot' as someone gets close to the hidden toy or 'cold' and 'freezing' as they move away.
- Encourage the children to make exciting dens for toys in your outside area. Hide a toy inside one and using a stopwatch or an egg timer see how long it takes the children to find it.

Learning objectives
- To talk about positions, directions and distances
- To measure short periods of time in simple ways
- To count steps and movements
- To identify numerals and place them in order
- To give and follow instructions and clues involving several ideas and actions
- To use what has been learned about media and materials in original ways, thinking about uses and purposes and building to own designs

Extensions/variations
- Hide some plastic and wooden numbers in your sand tray.
- Encourage children to find all of the numbers and then to put them all in the right order.

Links to home
- Suggest that the children play the same hide and seek game with their brothers and sisters or friends at home.

What to do
- Tell the children that you have hidden a teddy in the room and that you want them to find it. Say that you will give clues to the children such as 'move forward', 'move to your left' or 'move to your right' and 'reach up high'.

Weather chart

Learning objectives

- To collect and handle simple data
- To create a chart, select the right numerals to represent 1–10 and record the data
- To ask and answer questions and find out information from the chart
- To count entries and compare quantities, using the language 'more' and 'fewer'
- To talk about features of the environment and own observations and experiences of different types of weather

What to do

- Ask the children to think about the type of weather they really enjoy. Say that today they are going to make a chart which will show the most popular type of weather.
- Talk to them about symbols of weather. If they like hot weather they could use the sun symbol. If they like rain, they could use the cloud with rain drops coming out of it. If they like snow, they could use the snowman symbol.
- In front of them, draw an axis on the large piece of card with the types of weather along the horizontal axis and the number of children up the vertical axis.
- Encourage children to draw symbols to represent their favourite weathers (sun, rain, snow, wind, clouds, etc.), then stick them onto thin card, cut them out and put a sticky fastener on the back of each one.
- Ask the children to pick a symbol that describes the weather they like best and stick it on the chart.

Extension/variation

- Once the graph is complete, encourage the children to ask and answer questions, using the information, such as: *How many children like hot weather? How many like the snow? How many like the rain? Which is the most popular type of weather? Which is the least popular?*

Topic
Weather

Resources
- Large piece of card
- Sticky fasteners
- Black felt-tipped pen

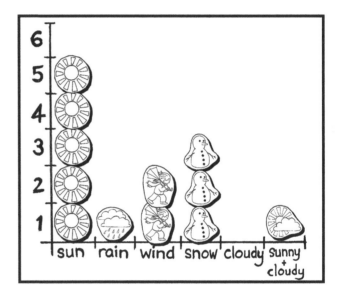

Links to home

- Ask the children to find out what their parents' favourite type of weather is, and perhaps what their favourite month is and why.

**Personal, Social and Emotional Development
with Understanding the World and Mathematics**

Animal homes

Topic
Animals

Resources
■ A variety of soft toys
■ Large and small boxes
■ Tubes of various sizes
■ Tissue paper
■ Sticky paper
■ Newspaper
■ Sticky paper
■ Material
■ Paints
■ Glue
■ Brown tape
■ Masking tape
■ Digital camera

⚠ Adult supervision will be needed when the children are using scissors.

Learning objectives
● To use familiar objects and common shapes to create and build representational models
● To begin to use mathematical names for solid 3D and 2D shapes
● To select tools and techniques needed to adapt shapes, assemble and join materials and cut them to size

What to do
● Present to the children a selection of soft animals and tell them that they need homes.
● Talk to them about what a home is and what an animal needs inside their home.
● Some children may suggest that it would need a bed or even an 'en suite' toilet.
● Encourage the children to look at the boxes and animals that you have collected.
● They need to think about which would be the right box for each animal. Ask the children to put the animals inside the boxes to check for size.

● Once the children have carefully selected their boxes and animals, have them work independently to cut tissue paper, material or newspaper to size for the carpet. They could paint the walls and use sticky paper to decorate them. Inside, they could use smaller boxes to create a bed or anything else that they have suggested.
● You might help the children to cut doors and windows to size in their boxes.
● As they work use appropriate vocabulary such as *cube*, *cuboid* and *cylinder*. Children love the 3D shape names. Encourage them to look for 2D shapes on 3D shapes. For example, you might say, 'Can you see any squares on this cube?

Extensions/variations
● Take a photograph of each child with their finished model (using a digital camera if possible). Invite the children to talk about their models and say why they like them.
● If possible, encourage the other children to ask questions such as 'How did you make it?'

Links to home
● Put a plea out to parents several weeks before this activity, asking for a variety of junk material.

Wrap it up

Learning objectives
● To select particular named shapes and use mathematical language to describe them
● To recognize, create and describe patterns
● To safely use and explore a variety of materials and techniques, experimenting with colour, design and texture

What to do
● Show the children a variety of pieces of wrapping paper and encourage them to identify the patterns, such as red balloon, green balloon, red balloon, green balloon.
● Ask them to name or describe some of the shapes on the paper, such as squares, stars, circles or triangles.
● Provide each child with a large piece of sugar paper.
● Encourage them to select two different sponges and two trays of paint.
● Help them to create one line of pattern across the top of the paper, such as red circle, green triangle, red circle, green triangle.
● Keep saying the pattern vocabulary to them as they work.
● Once they have done one line of the pattern, encourage them to start the next line and keep going until the whole sheet is covered.

Extensions/variations
● Once the sheets are dry, help the children to wrap their boxes.
● Encourage the children to count the shapes in their patterns and to say how many red circles they have used or how many green triangles, etc.
● If you have a house in your role-play area, you may decide to allow the children to print a repeating-pattern wallpaper.

Links to home
● Ask parents to donate large boxes and used sheets of wrapping paper.
● Ask the children to be pattern detectives in their own homes. They may be able to spot patterns on their curtains, shower curtains or wallpaper and look for shapes within the patterns.

Topic
Celebrations

Resources
■ Sheets of wrapping paper as example
■ Large pieces of sugar paper
■ Poster paints mixed with PVA glue to make it thicker and better for printing
■ Trays
■ Shaped sponges
■ Large boxes
■ Sticky tape
■ Different objects to go inside the boxes to make them light or heavy (place the boxes in your maths area so that the children can lift them, placing them in order by weight)

Sock shapes

Topic
Families

Resources
- A variety of socks
- A washing line at the children's height so that they can feel the socks
- 2D shapes
- 3D shapes
- Old box
- Wrapping paper
- Material and cord string to make a feely bag

Learning objectives
- To use mathematical names for 2D and 3D shapes
- To use everyday language to talk about size and capacity, to compare objects and to solve problems
- To describe properties of shapes in order to identify them

What to do
- Set up in front of the children a washing line, and peg on a variety of socks containing different shapes.
- Say to the children that you wonder what is inside the socks.
- Put your hand in, feel, and say, *Oh, I think it is a shape and it has three corners and three sides, I wonder if anyone can guess what shape I am feeling?* Help them towards the answer that is a triangle.
- Repeat for the next sock, describing the properties and inviting the children to guess what shape is inside.
- Ask a child to take on your role and describe the shape they are feeling.
- Display the socks in your setting, so that your children can go to them in spare minutes and play the game.

Extensions/variations
- As well as socks for this feeling activity, you could use a feely bag or a feely box. A feely bag is better if it has a drawstring at the top and perhaps a felt face on the front to motivate the children. A feely box can be made by covering an old box with bright wrapping paper and cutting a hole large enough to fit a hand through at the top.
- Instead of using 2D shapes, you could use 3D shapes.

Links to home
- Encourage parents to use mathematical language with their child to discuss shapes in their environment. For example: *same because*; *different because*; *curved*; *has corners* and *straight sides*.

© Mavis Brown and Rebecca Taylor
Brilliant Publications

Footsteps game

Learning objectives

- To measure distances in simple ways and with non-standard units
- To estimate numbers within measurements and check by counting them
- To count actions or objects which cannot be moved
- To move freely and with pleasure and confidence in a range of ways

Topic
Health

Resources
- Measuring tools such as rulers and tape measures
- Meter rulers
- Playground paint

What to do

- In your outside area, encourage the children to estimate how many jumps it will take them to get from one side to the other. Allow them to test it out and see if their estimation was close.
- Show them how to do giant steps. Encourage them to estimate how many they will need to do to get from one end of the playground to the other.
- Show them how to take baby steps and ask them to think about how many of these they will need to do to get across the playground.
- Go inside and repeat the activities but this time perhaps measure your corridor or big play room. Encourage them to tell their findings to their friends, for example *It takes me ten giant steps to get from one end of the corridor to the other.*

Extensions/variations

- If possible, paint a footprint trail in your outside area that the children can follow.
- Have a display of measurement tools such as rulers, meter rulers and tape measures because it is important for the children to know that they exist.
- As well as using their feet to measure show them how they can use their hands instead, trying out how many hand-spans they can measure along a table.

Links to home

- Ask children to find out how many baby steps and how many giant steps it takes for them to walk the whole length of their garden path at home. Suggest that they ask their parents to try and ask if they know why the numbers are different.

My shape house

Topic
Homes

Resources
- A4 sugar paper
- A large number of sticky shapes in a variety of sizes
- Glue
- Black felt-tipped pens

Learning objectives
- To show an interest in shape and space by playing with shapes to make pictures, patterns and arrangements
- To show an awareness of similarities in shapes in the environment
- To use shapes appropriately for tasks
- To use mathematical names for 2D shapes
- To select particular named shapes
- To manipulate materials to achieve a planned effect

What to do
- Gather the children in front of you and encourage them to talk about shapes by asking the following questions: *What shapes can they see in the room? What shape are the windows? Can they see anything else in the room which is the same shape as the window?*
- Provide each child with a piece of sugar paper and a selection of coloured sticky shapes of various sizes.
- Encourage the children to select shapes to make a picture of a house.
- Which shapes could they use to represent the roof? Which shape could they use for the doors? Which shape could they use for the path?
- Encourage the children to arrange their shapes before sticking them down.

Extensions/variations
- During another session, you may ask the children to make a shape person, asking questions such as: *Which shape could they use to represent the head?*
- Once they have finished arranging their picture and have stuck it down, provide them with a black felt-tipped pen to draw a face on the head. Rectangles make great pony tails!

Links to home
- Encourage parents to ensure that the children have access to lots of puzzles. Making puzzles is a great way to learn about fitting shapes together and can give a lot of satisfaction when they are completed. Many libraries have jigsaws to borrow.

Design a shape garden

Learning objectives
- To use mathematical names for 2D shapes confidently
- To show interest in shapes in the environment by talking about them, making arrangements and using shapes appropriately for tasks
- To select a particular named shape
- To use common shapes to create patterns and build models
- To manipulate materials to achieve a planned effect, experimenting with colour, design and texture

What to do
- Talk to the children about the shapes in your garden. Say that today you are going to design a garden using coloured shapes.
- Talk to them about moving the shapes around and only sticking them down when they are happy with their position.
- Encourage them to plan, design and create their own garden. Model ideas, such as a patio using squares, a flower bed using rectangles, circles for flowerpots and ponds, triangles for crazy paving and a path running through, by making your own design alongside them. Cut some shapes from different materials, such as fabrics and textured papers.
- Ask questions such as: *Which shape could we use for the shed?*, *Which shape would fit in this corner of the garden?*
- Show the children that by moving the shapes around two triangles can make a square.
- You could display your garden plans in a role-play garden centre, indoors or outside.

Extensions/variations
- The children may like to use the same strategy to design a bedroom or lounge.
- Once they have planned it with flat 2D shapes, perhaps they could use 3D junk materials to make a model of it.

Topic
Gardening

Resources
- Sugar paper
- A variety of coloured sticky shapes such as circles, triangles, rectangles and squares
- Glue
- Junk modelling materials
- Textured papers and fabrics

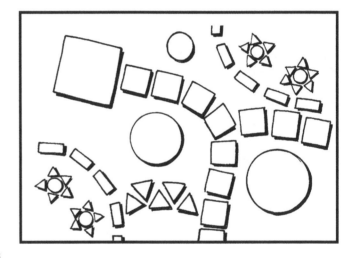

Links to home
- Encourage the children to talk about the shapes that they have in their garden. They might have stepping-stone circles going across their grass. Be sympathetic to each child's circumstances. If they do not have their own garden, they could talk about one belonging to a relative or friend.

Enormous sunflowers

Topic
Gardening

Resources
- Photographs and pictures of sunflowers
- Packets of sunflower seeds
- Magnifying glasses
- Watering cans
- Flowerpots
- Peat
- Compost
- An area where you can store the seedlings that is on the same height as the children
- Camera

Learning objectives
- To use mathematical language to describe size confidently
- To order items by height and compare their characteristics
- To count reliably with numbers 1–20
- To estimate a number of seeds and check by counting them

What to do
- This activity is best done in April.
- Ask the children to look at a packet of sunflower seeds and guess how many are inside.
- Open it up carefully and encourage the children to help you count out the seeds.
- Ask the children to use the magnifying glasses to observe the seeds
- Show the children how to plant a seed and then allow them to do the same independently.
- Plant a few extras as well, just in case some children's seeds do not grow.

- As the seedlings grow, encourage your children to move them around until they are in height order.

Extensions/variations
- Take photographs of the sunflowers as they progress so that you can have a lasting record. Encourage the children to look at their plants everyday to observe any changes and to decide whether they need water.
- After several weeks of growth, allow the children to take their seedlings home to plant. Plant your spare ones in your outside area so that the children can continue to observe the growth.

Links to home
- Put a plea out to parents for pots, sunflower seeds and compost. It is better if every child has a sunflower but this means you will need a lot of resources.

Personal, Social and Emotional Development with Understanding the World and Mathematics

© Mavis Brown and Rebecca Taylor
Brilliant Publications

Big animal, little animal

Learning objectives
- To use the language of size
- To compare and sort items by size
- To work cooperatively within a small group, sharing fairly and taking account of one another's ideas of how to organize the activity

What to do
- Talk to the children about the word *big*. Explain it means that you are talking about size. Can they name any big animals, for example *elephant* or *rhinoceros*?
- Can they name any little animals, for example *mouse*, *bird* and *ladybird*?
- Show the group of children your two large pieces of card. Write *big* on one and *little* on the other. Ask the children to look through the magazines for pictures of big and little animals. When they have found some they should cut them out and stick them on to the right piece of card.
- Encourage the children to talk to each other about where they are sticking their pictures and to share the equipment that they are using.
- When they have finished, encourage the children to recite the names of the animals.
- If the children can think of any other animal whose pictures are not in the magazine encourage them to draw them.

Extensions/variations
- You might encourage the children to reinforce their learning by repeating the activity but this time changing the theme to vehicles.
- Use size vocabulary in your setting. For example, *Taylor is going to sit on the big chair*, *Lauren is drinking from the little cup*.

Links to home
- Ask the children to find big and little objects at home and to count how many big objects and how many little objects they can find.

Topic
Animals

Resources
- 2 large pieces of card
- Glue
- Children's safety scissors
- A variety of magazines with animal pictures
- A variety of magazines with pictures of vehicles

How many cups?

Topic
Water

Resources
- Water tray
- Water
- Plastic containers
- Cups
- Sticky labels
- Plastic teapots

Learning objectives
- To use everyday language to talk about size and capacity, to compare quantities and objects and to solve practical problems
- To order containers by capacity
- To count reliably with numbers 1–20
- To compare sets, using the language of 'more' and 'fewer'

What to do
- Gather a group around the water tray and ask them to select two containers and one cup.
- Tell them that they are going to find out which container holds more by counting how many cups they can pour into each one.
- Fill a cup with water and show the children that it is full.
- Explain that every cup that they pour in must be full otherwise it will not be fair.
- Help the children find out how many cups it takes to fill each container and then decide which container holds more.

- Extend the learning by encouraging the children to order three containers.
- Sometimes the shape of a container can mislead you about the quantity it holds and it is good for the children to discover this.
- Older children could stick a sticky label on to containers, saying the number of cups they hold.

Extensions/variations
- Encourage children to have tea parties for their toys in the outside area. Help them to fill a plastic teapot with water and then find out how many cups it will fill.
- Ensure a plentiful supply of water is available to the children during the summer. Allow them to pour water into cups for everyone.

Links to home
- Ask parents to donate plastic bottles and containers for this activity. Remember glass bottles are not suitable in case they crack or smash.

Let's go to market

Learning objectives
- To begin to develop an understanding that numbers are labels
- To begin to add and subtract single digit numbers and count on or back to find answers, while engaged in practical activities
- To begin to recognize coins and to use everyday language to talk about money
- To show interest in different occupations and develop positive relationships with members of the community

What to do
- Send a letter home to parents, asking for permission to take their child to a shop. You could also ask for volunteers to help you look after the children, a ratio of 1 : 2 adults to children is excellent.
- Use the walk to the market or shop as a real learning experience for the children, pointing out house numbers, shop numbers and numbers on passing buses.
- At the market or green grocer encourage each child to make their purchase talking to them about the prices, the coins that they are using and the change that the green grocer gives.
- When the children get back to your setting, encourage them to describe the journey, talking about what they saw and any numbers that they spotted.
- Encourage the children to draw pictures of their trip which you could send to the green grocer as a thank you.

Extension/variation
- Set up a green grocer stall in your role-play area so that the children can re-enact what they did. Use plastic, papier mâché or real fruits and vegetables, provide scales, a till and card for the children to write prices on. If you are lucky, your green grocer might donate a pack of paper bags for the children to use.

Links to home
- Ask the parents to send in a small amount of money to enable their child to buy some fruit or vegetables. Perhaps they could buy three carrots for their family's tea.

Topic
Food and shopping

Resources
- Letter home to parents
- Carrier bags to help the children carry their items
- First aid kit
- Find a local green grocer or market which would be prepared to serve a large group of children. Many are more than willing to help, and see it as an excellent learning experience for the children
- Green grocer in role-play area: plastic, papier mâché or real fruits and vegetables; scales, till, play money etc.

 Adult supervision is required at a ratio of 1 : 2 adults to children.

Weigh the banana

Topic
Food and shopping

Resources
- A variety of heavy objects such as a door stop, a full suitcase, a thick book
- A variety of light objects such as a piece of paper, a feather, a paperclip, a thin book
- 2 hoops
- 2 pieces of card, one labelled **heavy** and the other labelled **light**
- A bunch of bananas
- Multi-link cubes
- Clear balancing scales

Learning objectives
- To use everyday language to talk about weight
- To understand that weight is not dependent upon size, but upon materials
- To order two objects by weight

What to do
- Gather the children together and invite them to sit in a circle. Place the two hoops in the middle. Place in one the **heavy** sign and in the other the **light** sign.
- Show the children a variety of objects and tell them that today you are going to find out if the objects are heavy or light. Ask them to suggest ways to do this. Help them see that a good way of seeing whether an object is heavy or light is to hold it in your hand.
- Do this for a few objects emphasizing when something is heavy.

- Choose two objects that have not yet been placed in either hoop and ask a child to hold one in each of her hands.
- Show her how to move her hands up and down to decide which object is heavier and which is lighter.
- Beware of some children saying nothing is heavy in order to appear strong!

Extensions/variations
- Present the weighing scales to the children. Put a banana in one pan. Ask a child to count cubes as they place them in the other side of the pan until the scales balance.

Links to home
- Encourage the children to find out if there are any scales in their house.
- They could ask their parents what they use scales for. Perhaps they weigh ingredients for cooking on kitchen scales or weigh themselves on bathroom scales.

Personal, Social and Emotional Development with Understanding the World and Mathematics

Autumn scales

Learning objectives

- To order two items by weight
- To estimate how many objects can be seen and check by counting them
- To count reliably with numbers 1–20
- To make observations of animals and plants, explain why some things occur and talk about changes

What to do

- This activity is best performed in the Autumn.
- Tell the children that you need their help to create a nature table. (See Links to home for their contribution.)
- Encourage them to think about what is special about autumn and how we recognize autumn by looking at the natural environment.
- Discuss which animals are usually seen in autumn and show the children examples of toy animals like squirrels, hedgehogs and foxes.
- Invite the children to help you to set up a nature table by laying out a green piece of material, adding wicker baskets to put the nature items and the soft toys in, and a pair of bucket scales.
- Show the children how to put an animal in one bucket and then count the conkers into the other bucket until they balance to find out how much the animal weighs. For example, 'Our hedgehog weighs 15 conkers.'

Extensions/variations

- Next, weigh the squirrel using the conkers. Ask the children to decide: *Is the squirrel heavier or lighter than the hedgehog?*
- Encourage the children to find out how many leaves the hedgehog weighs and to think about why it takes more leaves to balance the hedgehog compared with conkers?
- Suggest that children put a selection of conkers in their hands and ask friends to estimate the numbers of conkers Ask them to count together to see if the estimate was correct.

Links to home

Topic
Seasons

Resources
- Stuffed toy animals, for example a squirrel, hedgehog, fox and badger
- Green material
- Wicker baskets
- Bucket scales
- A variety of natural items collected by the children

⚠ Children should be reminded never to eat conkers or any wild berries as they could be poisonous.

- Encourage the children to visit local parks and woods with an adult to collect items of nature. Show them examples of what you would like them to collect, for example pine cones, conkers and leaves. Stress to them that they must never take these from branches but instead look for them after they have fallen.
- Remind the children not to eat the conkers as they are poisonous.

© Mavis Brown and Rebecca Taylor
Brilliant Publications

**Personal, Social and Emotional Development
with Understanding the World and Mathematics**

Who helps us?

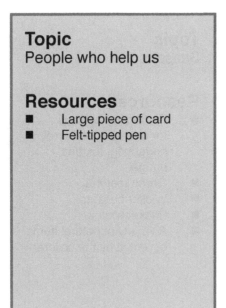

Topic
People who help us

Resources
- Large piece of card
- Felt-tipped pen

Learning objectives
- To use everyday language related to time
- To order and sequence familiar events
- To measure short periods of time in simple ways

What to do
- Gather the children together and say that today you are going to think about people who help them.
- Ask them to remember what they needed help with this morning. For example, somebody might have opened a door for them, or helped with fastening buttons or made them breakfast.
- Encourage them to help you make a list of all the people who help out in your setting, such as parent helpers, the dustmen, post person, milk delivery person. How many are there?
- Then sing the days of the week with the children. How many days are there?
- Write the days of the week on your easel and then get the children to think about which day people help. For example, you might write that Ben's mum helps on a Wednesday.

Extension/variation
- Throughout the year, the children could help you thank your helpers by making cards and pictures for them. You might decide to do this at Easter and Christmas but also perhaps after this activity because it will really reinforce to the children just how many people help them. *How many cards will they need?*

Links to home
- Encourage the children to think about all the people who help them at home. They might have a cleaner, a post person, a newspaper delivery person, dustmen, pizza delivery person and gardener!

Let's be robots

Learning objectives
- To talk about positions, directions and distances
- To use positional language
- To count steps and movements
- To select and use technology, such as a remote-controlled or programmable toy
- To give and follow instructions involving several ideas and actions

What to do
- You will find that nearly all the children in your setting have a programmable toy and will love this activity.
- You may have a selection of remote-control toys or you may need to ask the children to bring in some of their own.
- Invite the children to sit in a circle and place the toys in the middle. Encourage the children to name them and suggest how they work.
- Ask questions to activate a discussion: *What do you have to do to make them work?*; *Do they need batteries to work?*; *Do you have to press an 'on' button?*; *How do you make them move forwards?*; *Do you have to press the arrow?*
- Ask some of the children to programme a toy to get to a certain place in your setting such as the door or wall.
- Tell the children that you are now a robot. Use a robot voice and become very robotic. Ask the children to give you instructions.

Extensions/variations
- Help the children to extend their instructions by using vocabulary such as *two steps forward*, *two steps back*, *turn to the right* and *turn to the left*.
- Swap over and let the children be robots.
- Remind them to listen to your instructions. Provide junk modelling materials and craft pieces and support the children as they make their own robots. Cover them with foil and use plastic bottle tops as eyes and nose.

Links to home
- Suggest that the children play robots with their parents at home and think of what jobs

Topic
Shapes

Resources
- A variety of programmable toys such as a remote- control car, boat, dog or robot
- A variety of boxes and tubes
- Tubes
- Plastic bottle tops, foil and paint

they would programme them to do if they did have robots in their houses. Perhaps tidying their room would be the first job.

Teddy bear clock

Topic
Seasons

Resources
- Teddy bear clock template on page 202
- White card
- A variety of clocks
- Wax crayons
- Scissors
- Split pins
- Black felt-tipped pens
- Templates of numbers

⚠️ Caution needs to be applied when attaching the split pin.

Learning objectives
- To use everyday language to talk about time
- To order and sequence familiar events
- To understand that there are specific moments in time at which things happen
- To recognize numbers on a clock face

Preparation
- Photocopy on to white card the template on page 202. Allow one for each child, for them to colour in later.

What to do
- Show the children a variety of clocks.
- Talk about what time it is now, what time they might have a story and what time they will have their lunch.
- Tell the children that they are going to make a teddy bear clock today.

- Provide the children with some wax crayons and encourage them to decorate their teddy.
- Laminate each if you have time. Cut out the teddy bears and the hands.
- Using a split pin, fix the hands to the middle of the clock.

Extensions/variations
- Invite the children to sit together with their clocks. Use a teaching clock or make a big clock from white card, with cardboard numbers and hands, and staple it to the wall.
- Make a time on your clock that is significant to the children, such as the time they arrive at or leave the setting. Ask if they can make the same time on their clocks.

Links to home
- Ask the parents to lend different types of clocks, for example alarm clocks, egg timers, travel clocks, wrist watches, sports timers, and make a display of them in your setting.
- Ask the children to count how many clocks they have in their home.

Personal, Social and Emotional Development with Understanding the World and Mathematics

© Mavis Brown and Rebecca Taylor
Brilliant Publications

Making a piece of furniture

Learning objectives

- To show an interest in shape and space by sustained construction activity and making patterns
- To show an awareness of similarities of shapes in the environment
- To use shapes appropriately for tasks
- To use mathematical names for 3D shapes
- To see how a 3D shape collapses into a net
- To make a piece of furniture that is the right size for a toy
- To manipulate and change 3D shapes
- To construct with a purpose in mind, using a variety of resources
- To safely use and explore a variety of materials, tools and techniques, experimenting with colour, design, texture, form and function

Topic
Homes

Resources
- A variety of junk modelling materials: boxes, tubes and yoghurt pots
- Poster paints
- Masking tape
- Toys

What to do

- Encourage the children to think about all the furniture that they have in their house.
- Tell them that they are going to make a piece of furniture for their favourite small toy.
- Give them many ideas, for example two cereal boxes and four tubes make a great bunk bed, or two cereal boxes and two smaller boxes all fitted together make a great armchair.
- Show the children how if you run your finger along the flaps of a cereal box and open up the ends, you can flatten it out to make a net. You can then rebuild the box, using masking tape along the edges ensuring that the grey side is on the outside making it easier to paint.
- Encourage the children to tell you what they are going to make before they get started.
- Then help them find the appropriate boxes and turn them inside out. Work with the children to help them join their boxes and tubes to make their planned model.

Links to home

- Encourage the children to a do a furniture count in their own houses and to find out how many chairs, tables, beds and sofas they have at home.

Extension/variation

- Encourage the children to paint their furniture, perhaps with a repeating pattern. They might want to put squares of material on their models for cushions or tablecloths.

**Personal, Social and Emotional Development
with Understanding the World and Mathematics**

House shape game

Topics
Homes/Shapes

Resources
- Templates on pages 203–204
- Cord
- Felt-tipped pens
- Clear laminate sheets
- Scissors
- Plastic folder with zip
- Glue

Learning objectives
- To show an interest in shape and space by playing with shapes and making pictures, patterns and arrangements
- To use mathematical names for 2D shapes and talk about their properties
- To show an awareness of similarities of shapes in the environment
- To select particular named shapes
- To count reliably with numbers 1–10
- To play cooperatively, taking turns and sharing fairly
- To work as part of a group, understanding and following the rules

Preparation
- Photocopy on to card six times the template on page 203. Colour with felt tipped pens if you want to. Laminate each sheet and then cut out the corresponding shapes that will fit on the base board. Photocopy on to thin card the die template on page 204, colour it in and laminate it to make it last longer. Store the bits for the game in a plastic folder with a zip.

What to do
- Invite six children to sit around a table or on the floor to play a game. Give them each a base board and lay out all the corresponding shapes around them.
- Encourage the children to throw the die and say what the shape is, collect the correct card and place it on their base board.
- Encourage the children to take turns and not shout out what the die has landed on so that everyone has a fair chance.
- The child who collects all their shapes first is the winner.
- Once you have got winner number one, you may like to carry on the game and have winner number two and winner number three.

Extension/variation
- Ask the children if they can see any symmetry on the house and explain what this concept means.

Links to home
- Ask the children to find out the answers to the following questions: *How many windows does their house have? How many doors? How many rooms? How many stairs?*

Peepo

Learning objectives
- To use mathematical names for solid 3D shapes and flat 2D shapes
- To learn mathematical terms to describe shapes
- To show interest in shape by talking about and comparing shapes and arrangements

What to do
- Show the children your examples of 2D shapes and encourage them to name them. Ask how many sides they have and how many corners.
- Ask the children to close their eyes. Put one of the 2D shapes into a big brown envelope. Very slowly, begin to reveal it. Ask the children to guess what the shape is, predicting from what they can see. If a corner is showing it might be a square or a triangle. If no corners are showing it might be a circle. When the whole shape is revealed encourage the children to shout 'Peepo'.
- If a child wants to, you could let them reveal the shapes while you sit with the other children to help them guess.
- Show the children the 3D shapes and ask if they can name any of them. From behind your easel, slowly reveal a part of the 3D shape and encourage them to predict the shape from what they can see. When the whole shape is revealed encourage the children to shout 'Peepo'.

Extension/variation
- Help the children to make their very own 'Lift-the-flap shape book'. Encourage them to stick a sticky shape on to each page and then place a piece of white paper big enough to cover the shape over the top. Fix the paper with a piece of clear sticky tape at the top so that it is a flap. Repeat for all the pages.

Topic
Shapes

Resources
- Cardboard 2D shapes
- Brown envelope or clothes-peg bag
- Plastic or wooden 3D shapes
- Easel
- A4 paper to make a book
- Sticky shapes
- Glue
- Clear sticky tape

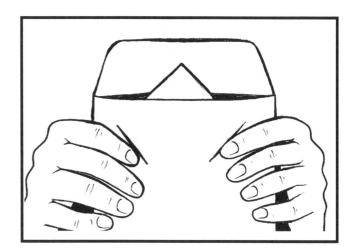

Links to home
- Encourage the children to take their books home and share them with their parents.
- Tell them that they should shout 'Peepo!' when they reveal the shape.

Harvest basket template

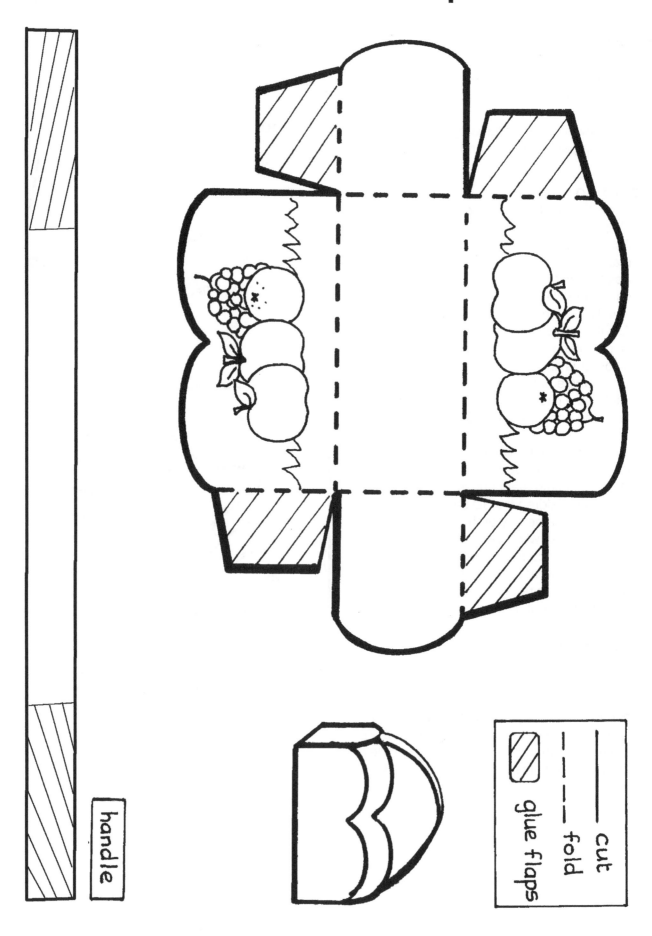

handle

cut
fold
glue flaps

**Personal, Social and Emotional Development
with Understanding the World and Mathematics**

Countdown to Christmas template

Bear washing line template

Personal, Social and Emotional Development with Understanding the World and Mathematics

Ten brown bears template

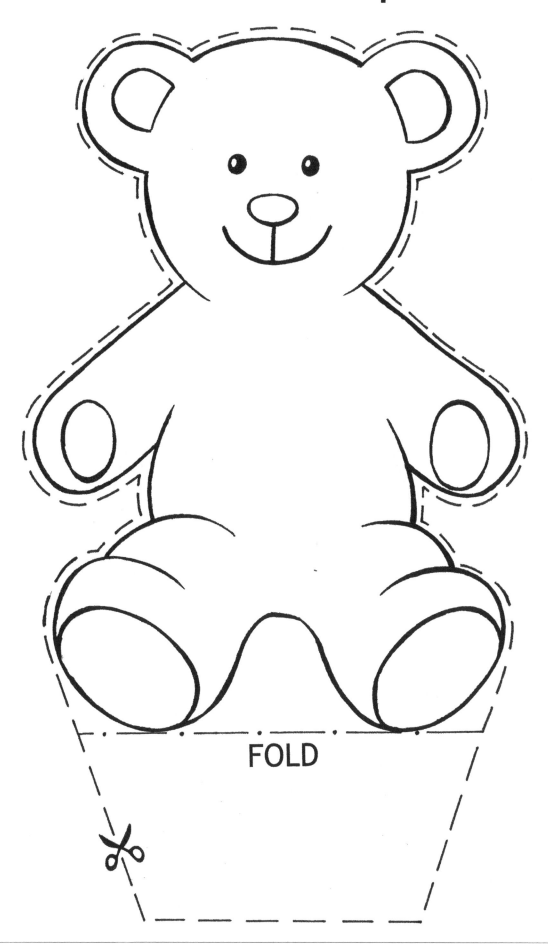

FOLD

Body bingo template 1

**Personal, Social and Emotional Development
with Understanding the World and Mathematics**

Body bingo template 2

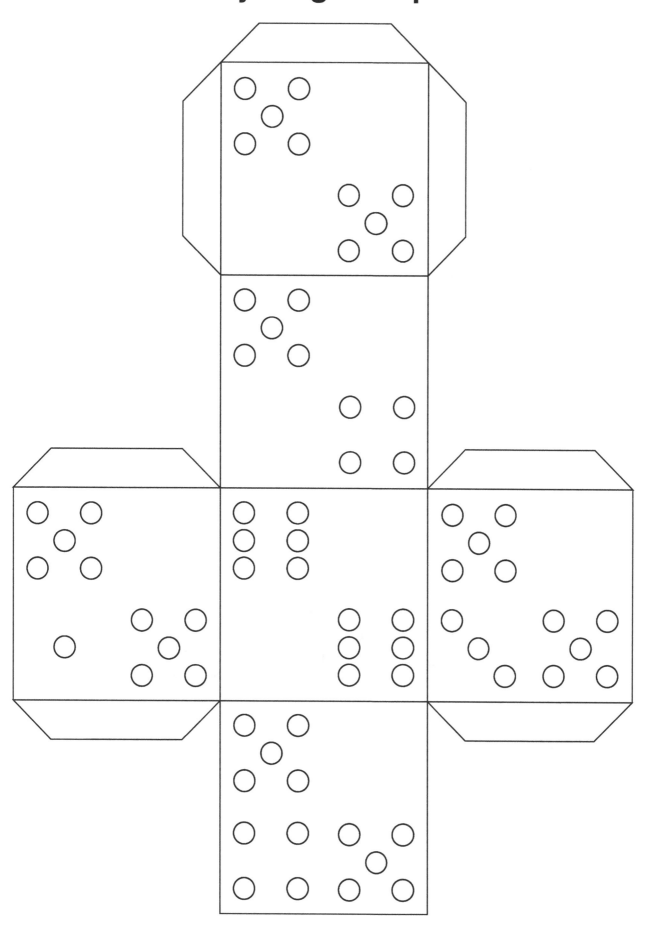

Bunny boxes template

 Personal, Social and Emotional Development with Understanding the World and Mathematics

Colourful cube game template 1

Colourful cube game template 2

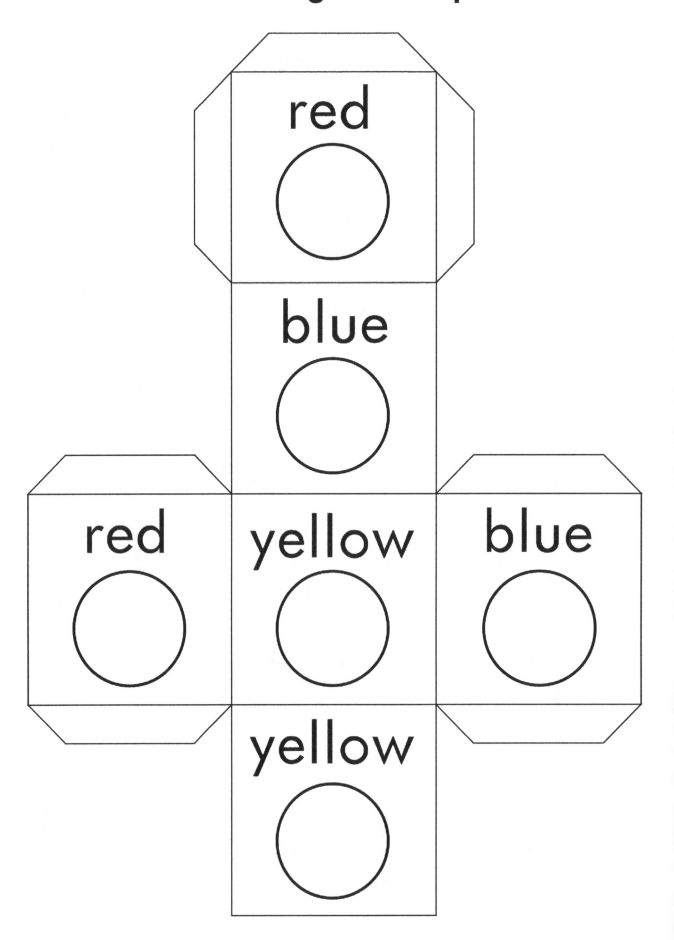

**Personal, Social and Emotional Development
with Understanding the World and Mathematics**

© Mavis Brown and Rebecca Taylor
and Brilliant Publications

My family number game template 1

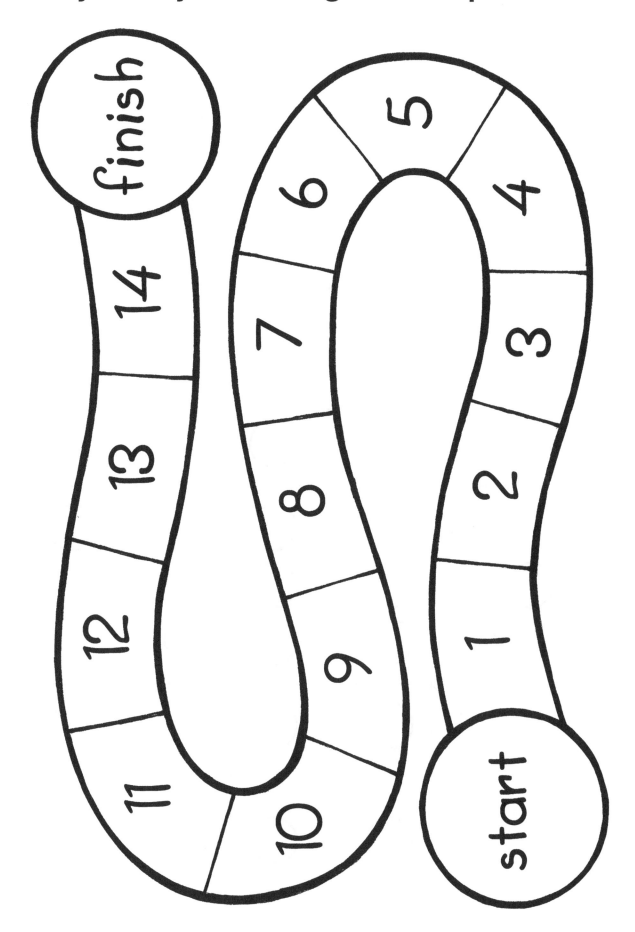

My family number game template 2

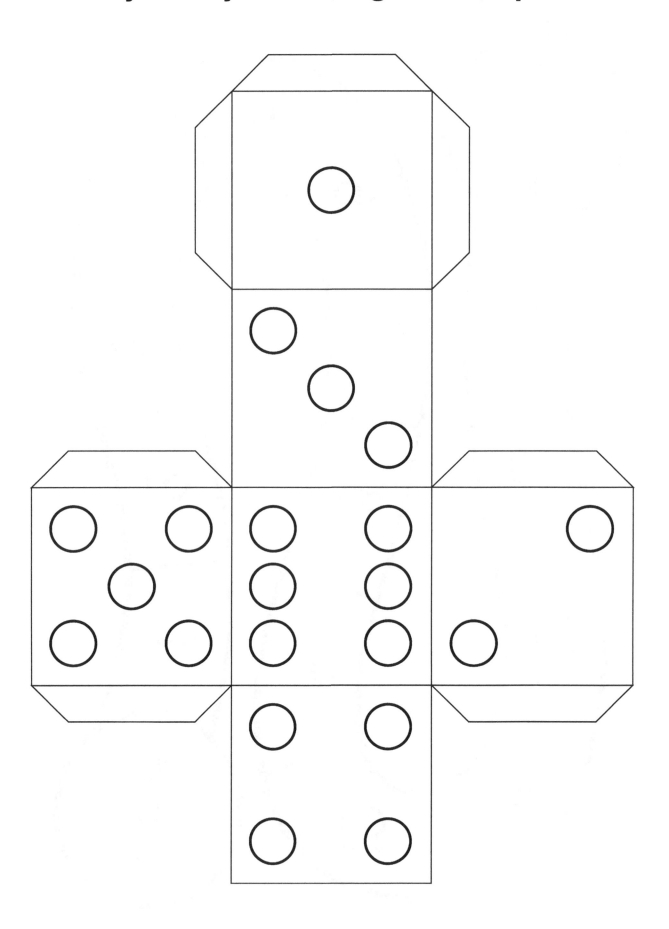

Family journeys template

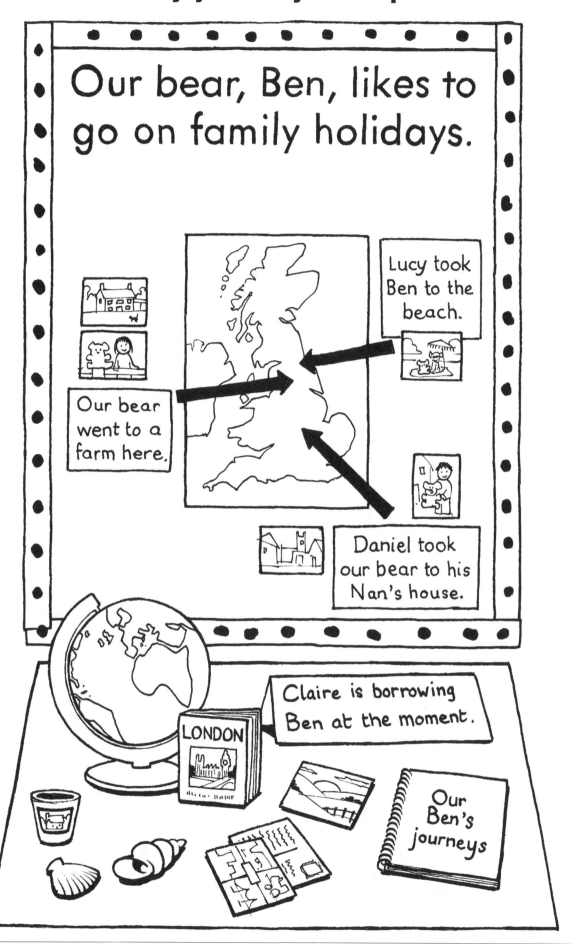

Our bear, Ben, likes to go on family holidays.

Lucy took Ben to the beach.

Our bear went to a farm here.

Daniel took our bear to his Nan's house.

Claire is borrowing Ben at the moment.

LONDON

Our Ben's journeys

Teddy bear clock template

Personal, Social and Emotional Development with Understanding the World and Mathematics

House shape game template 1

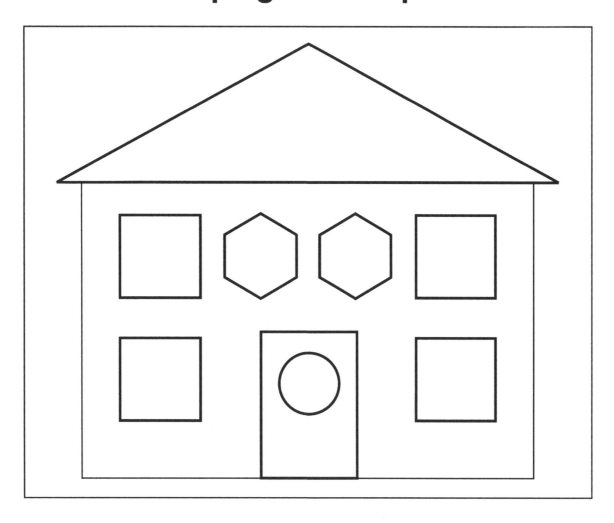

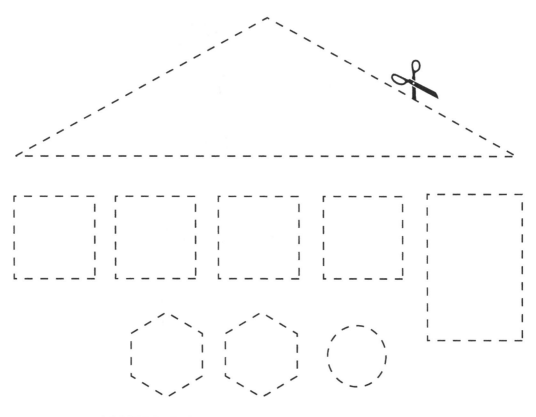

House shape game template 2

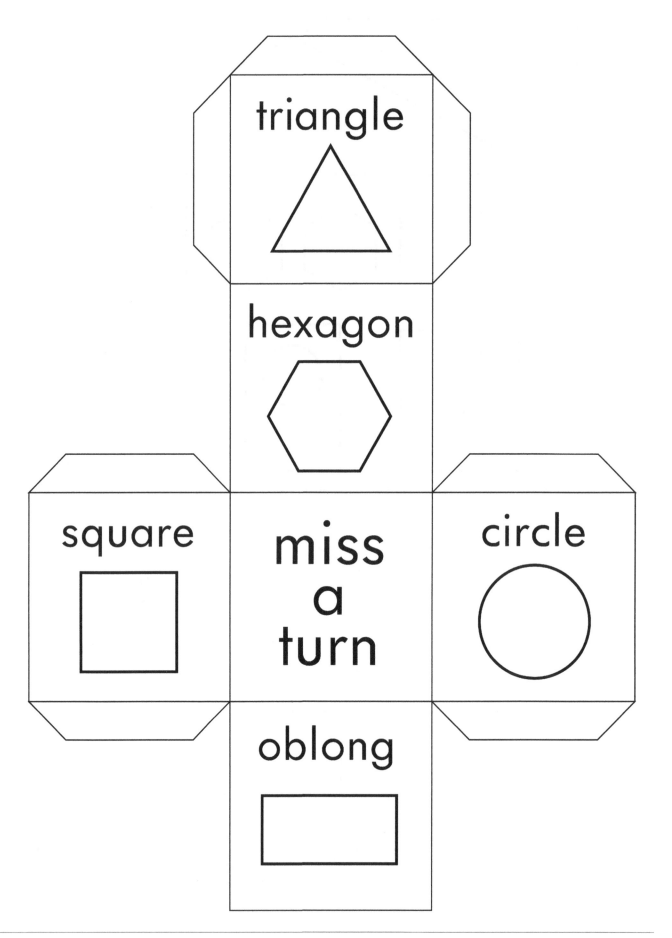

**Personal, Social and Emotional Development
with Understanding the World and Mathematics**

Characteristics of effective learning

Throughout all activities, practitioners need to be aware of the four themes of the Early Years Foundation Stage. Every child must be considered a *unique child* and given opportunities to form *positive relationships* within an enabling environment in order to make progress in *learning and development* in each of the prime and specific areas.

In addition to the *Learning Objectives and Early Learning Goals* listed for each activity, practitioners will seek always to encourage and observe the following *characteristics of effective learning*.

Playing and exploring

Children will find out and explore by showing curiosity and developing particular interests, using their senses and engaging in activities. They will play with what they know, acting out their own experiences through pretending and imaginative role-play. They should be willing to 'have a go' by seeking new and challenging activities and being confident to try things out.

Practitioners should join in with children's play without taking over, helping, supporting and modelling ideas, challenges and risk taking and setting an example that effort and practice improves and mistakes can be learned from.

Children need flexible and stimulating resources provided within calm and ordered indoor and outdoor spaces and uninterrupted periods of time to play and explore.

Active learning

Children will learn to become involved and concentrate on their chosen activities, displaying high levels of fascination and attention to details, maintaining focus and ignoring minor distractions. They need to develop perseverance, effort and persistence to 'keep on trying' through challenges and believe that a solution to a problem may be found or accept that an idea may not work out exactly as planned. They should enjoy meeting their own goals and challenges and be proud of their own accomplishments and achievements, without relying heavily on external praise or rewards.

Practitioners should support children in choosing their own activities, methods, plans and goals and talk with them about progress, challenges and successes. They may be encouraged to work together and learn from each other when appropriate. Specific praise for particular efforts, persistence, problem solving, good ideas and new skills acquired will help children to develop their own motivations.

Children should be provided with new and unusual activities that are linked to their current interests and given enough time and freedom for all to contribute and to become deeply involved.

Creating and thinking critically

Children will have their own ideas, think of new ways to do things and find ways of solving problems for themselves. As they make connections and notice patterns and sequences within their experiences, they will learn to make predictions, test and develop their ideas and understand cause and effect. They will then be able to make informed decisions and plans, check and change their strategies as they work and play and eventually review their approaches and activities.

Practitioners should model thinking aloud, describing problems, remembering previous experiences, making connections, finding out and trying different ideas and approaches. If children's interests and conversations are supported, and sustained shared thinking is offered when appropriate, they will learn to use the 'plan-do-review' process effectively.

Children should always engage in activities in order to find their own ways to represent and develop their own ideas, using techniques and processes that they may learn from others. Routines should be recognizable and understandable to both children and adults, but also flexible enough to ensure that both security and independent development are available within the learning community.

Table of learning opportunities

Activity	Page no.	Listening and attention	Understanding	Speaking	Moving and handling	Health and self-care	Self-confidence and self-awareness	Managing feelings and behaviour	Making relationships	Reading	Writing	Numbers	Shape, space and measures	People and communities	The world	Technology	Exploring and using media and materials	Being imaginative
First visits	18						✓											
Babysitter	19						✓											
Did you enjoy that?	20						✓	✓										
Colour matching	21						✓										✓	
What did you do at the weekend?	22						✓							✓				
I'm here	23						✓		✓									
Honey cake	24			✓			✓											
I'm a builder	25						✓	✓						✓			✓	
Sports day	26				✓		✓											
I want to be …	27						✓							✓				
Stranger danger	28					✓	✓											
Circle time	29						✓		✓									
Straw shapes	30				✓		✓											
The tree house	32						✓	✓									✓	✓
Underground	33						✓	✓										✓
Allsorts	34							✓										

Area	Aspect	35	36	37	38	39	40	41	42	43	44	45	46	47	48	49	50
Expressive Arts and Design	Being imaginative							✓									
	Exploring and using media and materials																
Understanding the World	Technology																
	The world						✓										
	People and communities		✓				✓		✓					✓			
Maths	Shape, space and measures																
	Numbers											✓					
Literacy	Writing																
	Reading																
Personal, Social and Emotional Development	Making relationships			✓		✓						✓					
	Managing feelings and behaviour	✓		✓	✓	✓	✓	✓	✓	✓	✓	✓	✓	✓	✓	✓	✓
	Self-confidence and self-awareness				✓												
Physical Development	Health and self-care								✓	✓							
	Moving and handling														✓		
Communication and Language	Speaking																
	Understanding															✓	
	Listening and attention																

Page no.	Activity
35	Can I have a biscuit, please?
36	Stepping stones
37	Laughing
38	Well done
39	New friends
40	Not everyone likes fireworks
41	Stop, look, listen
42	Potato latkes
43	Don't do that
44	It's mine
45	Me first
46	Nightmares
47	Keep the noise down
48	Angry
49	Time to stop playing
50	Don't swing on the curtains

**Personal, Social and Emotional Development
with Understanding the World and Mathematics**

Area	Aspect	51 Graffiti	52 Only one sweet wrapper	53 Sorry	54 Whoops-a-daisy!	55 Jamela's dress	56 The four Cs	57 Can't play outside	59 Tea time	60 At the café	61 Let me introduce myself	62 My friend	63 All the fun of the fair	64 Welcome to my nursery	65 Hello	66 Making a secret garden	67 Harvest basket
Expressive Arts and Design	Being imaginative								✓	✓				✓			
	Exploring and using media and materials															✓	✓
Understanding the World	Technology									✓							
	The world	✓	✓														
	People and communities									✓							
Mathematics	Shape, space and measures																
	Numbers																
Literacy	Writing																
	Reading																
Personal, Social and Emotional Development	Making relationships				✓				✓	✓	✓	✓	✓	✓	✓	✓	✓
	Managing feelings and behaviour	✓	✓	✓	✓	✓	✓	✓									
	Self-confidence and self-awareness										✓		✓	✓	✓		
Physical Development	Health and self-care																
	Moving and handling								✓			✓	✓				
Communication and Language	Speaking																
	Understanding																
	Listening and attention			✓													

Area	Aspect	68	69	70	71	72	73	74	76	77	78	79	80	81	82	83	84
Expressive Arts and Design	Being imaginative		✓								✓						
Expressive Arts and Design	Exploring and using media and materials																✓
Understanding the World	Technology																✓
Understanding the World	The world								✓		✓				✓	✓	
Understanding the World	People and communities		✓						✓	✓	✓	✓		✓	✓	✓	✓
Mathematics	Shape, space and measures								✓	✓					✓	✓	
Mathematics	Numbers																
Literacy	Writing				✓												
Literacy	Reading																
Personal, Social and Emotional Development	Making relationships	✓	✓	✓	✓	✓	✓	✓									
Personal, Social and Emotional Development	Managing feelings and behaviour		✓			✓	✓										
Personal, Social and Emotional Development	Self-confidence and self-awareness					✓						✓					
Physical Development	Health and self-care																
Physical Development	Moving and handling					✓											
Communication and Language	Speaking													✓			
Communication and Language	Understanding	✓															
Communication and Language	Listening and attention	✓								✓							

Page no.	Activity
68	Take a message
69	How thoughtful
70	Be kind
71	Thank you
72	Ring games
73	Round and round
74	Why won't you play with me?
76	All grown up
77	Religious stories
78	Festival of light
79	Happy Chinese New Year
80	This is me
81	Who's that knocking at my door?
82	Older and older
83	When I grow up
84	Let me entertain you!

Area	Aspect	85	86	87	88	89	90	91	92	93	94	95	96	98	99
Expressive Arts and Design	Being imaginative				✓			✓				✓			
	Exploring and using media and materials					✓									✓
Understanding the World	Technology							✓							
	The world		✓						✓	✓	✓	✓		✓	✓
	People and communities	✓	✓	✓	✓	✓	✓	✓	✓	✓	✓	✓	✓		
Mathematics	Shape, space and measures	✓							✓						
	Numbers														
Literacy	Writing														
	Reading														
Personal, Social and Emotional Development	Making relationships						✓					✓			
	Managing feelings and behaviour														
	Self-confidence and self-awareness														
Physical Development	Health and self-care														
	Moving and handling														
Communication and Language	Speaking													✓	
	Understanding														
	Listening and attention														
Page no.		85	86	87	88	89	90	91	92	93	94	95	96	98	99
Activity		Transport in days gone by	Where do your relatives live?	A day to remember	Our baby	Burning brightly	Where do you come from?	Help me to cross the road	Victorian homes	My home	Where do you work?	My painted house	Other children's toys	What does it feel like?	Balancing sandy shapes

**Personal, Social and Emotional Development
with Understanding the World and Mathematics**

Area	Aspect	100 Do plants drink water?	101 Make a rainbow	102 Where did the puddle go?	103 Seeing clearly	104 Let's get mixing	105 Beans means...	106 Who dropped that?	107 Mirror, mirror, on the wall	108 The power of water	109 Good morning	110 Where can we play?	111 Mini-beast hunt	112 Boil an egg	113 Overflowing with water	114 Icebergs	115 The power of air
Expressive Arts and Design	Being imaginative		✓														
	Exploring and using media and materials									✓							✓
Understanding the World	Technology																
	The world	✓	✓	✓	✓	✓	✓	✓	✓	✓	✓	✓	✓	✓		✓	✓
	People and communities																
Mathematics	Shape, space and measures														✓		
	Numbers																
Literacy	Writing																
	Reading																
Personal, Social and Emotional Development	Making relationships																
	Managing feelings and behaviour																
	Self-confidence and self-awareness																
Physical Development	Health and self-care											✓					
	Moving and handling																
Communication and Language	Speaking																
	Understanding																
	Listening and attention																

**Personal, Social and Emotional Development
with Understanding the World and Mathematics**

Area	Sub-area	Changes 116	Growing plants 117	Bouncing balls 118	Faster and faster 119	Where do animals live? 120	Where does rain come from? 121	Attractive toys 122	Flying high 123	Wrapping paper for a present 125	Quack 126	All change 127	Across space 128	Apples in this pie 129	What a picture! 130	Wibble, wobble 131	Happy festival! 132	Who is that? 133
Expressive Arts and Design	Being imaginative																✔	
	Exploring and using media and materials			✔					✔	✔				✔				
Understanding the World	Technology									✔		✔	✔	✔	✔	✔	✔	✔
	The world	✔	✔	✔	✔	✔	✔	✔	✔			✔		✔		✔		✔
	People and communities														✔			✔
Mathematics	Shape, space and measures			✔	✔													
	Numbers																	
Literacy	Writing	✔																
	Reading															✔	✔	
Personal, Social and Emotional Development	Making relationships										✔							
	Managing feelings and behaviour																	
	Self-confidence and self-awareness																	
Physical Development	Health and self-care																	
	Moving and handling																	
Communication and Language	Speaking	✔																
	Understanding																	
	Listening and attention																	

		134 Does it need batteries?	135 Let's find out about animals	136 Bars and bills	137 Switch on	138 Put out the fire	139 A letter to Santa	140 The lights have gone out!	141 All about the lion dance	144 Animals in sand and water	145 Colour towers	146 Animal hospital	147 Birthday shop	148 Countdown to Christmas
Expressive Arts and Design	Being imaginative								✓			✓		
	Exploring and using media and materials													
Understanding the World	Technology	✓	✓	✓	✓	✓	✓	✓	✓					
	The world	✓												
	People and communities								✓				✓	✓
Mathematics	Shape, space and measures										✓	✓	✓	✓
	Numbers									✓	✓	✓	✓	✓
Literacy	Writing						✓							
	Reading						✓							
Personal, Social and Emotional Development	Making relationships					✓				✓				
	Managing feelings and behaviour													
	Self-confidence and self-awareness													
Physical Development	Health and self-care													
	Moving and handling										✓			
Communication and Language	Speaking													
	Understanding													
	Listening and attention													

Personal, Social and Emotional Development with Understanding the World and Mathematics

EYFS Area	Aspect	149 Five little Diwali lamps	150 Number igloo	151 Number values	152 Bear washing line	153 Ten brown bears	154 Pass the animal	155 Let's get physical	156 Body bingo	157 Gingerbread men	158 Bunny boxes	159 Colourful cube game	160 We have a pet tiger	161 Our tigers keep eating numbers!	162 Combining groups of flowers	163 My family number game	164 My number book
Expressive Arts and Design	Being imaginative					✓											
	Exploring and using media and materials	✓				✓					✓						
Understanding the World	Technology																
	The world			✓													
	People and communities	✓									✓						
Mathematics	Shape, space and measures	✓						✓	✓	✓	✓	✓				✓	
	Numbers	✓	✓	✓	✓	✓	✓	✓	✓	✓	✓	✓	✓	✓		✓	✓
Literacy	Writing																
	Reading																
Personal, Social and Emotional Development	Making relationships						✓		✓			✓				✓	
	Managing feelings and behaviour																
	Self-confidence and self-awareness													✓			
Physical Development	Health and self-care									✓							
	Moving and handling			✓			✓			✓							✓
Communication and Language	Speaking																
	Understanding					✓			✓								
	Listening and attention												✓				

**Personal, Social and Emotional Development
with Understanding the World and Mathematics**

Area	Aspect	166 Family journeys	167 Family photographs	168 Shape collages	169 Setting up a toy shop	170 Hide and seek	171 Weather chart	172 Animal homes	173 Wrap it up	174 Sock shapes	175 Footsteps game	176 My shape house	177 Design a shape garden	178 Enormous sunflowers	179 Big animal, little animal	180 How many cups?	181 Let's go to market
Expressive Arts and Design	Being imaginative					✓											
	Exploring and using media and materials			✓				✓	✓			✓	✓				
Understanding the World	Technology																
	The world	✓					✓							✓			
	People and communities																✓
Mathematics	Shape, space and measures	✓	✓	✓	✓	✓	✓	✓	✓	✓	✓	✓	✓	✓	✓	✓	✓
	Numbers	✓			✓	✓	✓				✓			✓	✓	✓	✓
Literacy	Writing				✓												
	Reading				✓												
Personal, Social and Emotional Development	Making relationships													✓			
	Managing feelings and behaviour																
	Self-confidence and self-awareness																
Physical Development	Health and self-care																
	Moving and handling					✓											
Communication and Language	Speaking		✓		✓												
	Understanding					✓											
	Listening and attention																

Personal, Social and Emotional Development with Understanding the World and Mathematics

Category	Sub-area	Weigh the banana (182)	Autumn scales (183)	Who helps us? (184)	Let's be robots (185)	Teddy bear clock (186)	Making a piece of furniture (187)	House shape game (188)	Peepo (189)
Expressive Arts and Design	Being imaginative								
	Exploring and using media and materials						✓		
Understanding the World	Technology				✓				
	The world		✓						
	People and communities			✓					
Mathematics	Shape, space and measures	✓	✓	✓	✓	✓	✓	✓	✓
	Numbers		✓	✓	✓	✓		✓	
Literacy	Writing								
	Reading								
Personal, Social and Emotional Development	Making relationships							✓	
	Managing feelings and behaviour								
	Self-confidence and self-awareness								
Physical Development	Health and self-care								
	Moving and handling								
Communication and Language	Speaking								
	Understanding		✓						
	Listening and attention								

Index of topics

**Personal, Social and Emotional Development
with Understanding the World and Mathematics**

Lightning Source UK Ltd.
Milton Keynes UK
UKOW07f1105240216

269000UK00004B/23/P